# About Island Press

Since 1984, the nonprofit Island Press has been stimulating, shaping, and communicating the ideas that are essential for solving environmental problems worldwide. With more than 800 titles in print and some 40 new releases each year, we are the nation's leading publisher on environmental issues. We identify innovative thinkers and emerging trends in the environmental field. We work with world-renowned experts and authors to develop cross-disciplinary solutions to environmental challenges.

Island Press designs and implements coordinated book publication campaigns in order to communicate our critical messages in print, in person, and online using the latest technologies, programs, and the media. Our goal: to reach targeted audiences—scientists, policymakers, environmental advocates, the media, and concerned citizens—who can and will take action to protect the plants and animals that enrich our world, the ecosystems we need to survive, the water we drink, and the air we breathe.

Island Press gratefully acknowledges the support of its work by the Agua Fund, Inc., Annenberg Foundation, The Christensen Fund, The Nathan Cummings Foundation, The Geraldine R. Dodge Foundation, Doris Duke Charitable Foundation, The Educational Foundation of America, Betsy and Jesse Fink Foundation, The William and Flora Hewlett Foundation, The Kendeda Fund, The Andrew W. Mellon Foundation, The Curtis and Edith Munson Foundation, Oak Foundation, The Overbrook Foundation, the David and Lucile Packard Foundation, The Summit Fund of Washington, Trust for Architectural Easements, Wallace Global Fund, The Winslow Foundation, and other generous donors.

CLIMATE SAVVY

# *Climate Savvy*

ADAPTING CONSERVATION AND RESOURCE MANAGEMENT TO A CHANGING WORLD

Lara J. Hansen and Jennifer R. Hoffman

ISLANDPRESS
Washington | Covelo | London

Library of Congress Cataloging-in-Publication Data

Hansen, Lara J.
Climate savvy : adapting conservation and resource management to a changing world / Lara J. Hansen and Jennifer R. Hoffman.
p. cm.
Includes bibliographical references and index.
ISBN-13: 978-1-59726-685-7 (cloth : alk. paper)
ISBN-10: 1-59726-685-X (cloth : alk. paper)
ISBN-13: 978-1-59726-686-4 (pbk. : alk. paper)
ISBN-10: 1-59726-686-8 (pbk. : alk. paper)
1. Conservation of natural resources. 2. Natural resources—Management. 3. Climatic changes.
I. Hoffman, Jennifer R. II. Title.
S936.H36 2010
304.2'5—dc22
2010035547



Printed on recycled, acid-free paper

Manufactured in the United States of America

10 9 8 7 6 5 4 3 2 1

KEYWORDS: Climate change, adaptation, resource management, resilience, restoration, invasive species, conservation, natural resources, connectivity, climate models, protected areas, changing world, natural resource policy, government agencies, conservation biology, restoration ecology, climate scientists, endangered species, assisted migration, vulnerability assessment.

CONTENTS

# ACKNOWLEDGMENTS

This book would never have seen the light of day had we not received the support and encouragement of a multitude of people.

The following individuals generously reviewed one or more chapters of the book: Tom Lovejoy, Hannah Reid, Hector Galbraith, John Matthews, Emma Tompkins, Earl Saxon, Erica Fleischman, Caroly Shumway, Anton Seimon, Marianne Fish, Gary Tabor, Lance Morgan, Marah Hardt, Robie MacDonald, Craig Segall, Jeb Byers, Joanna Ellison, Richard Hobbs, Dominique Bachelet, Katharine Hayhoe, Teresa A'mar, Diane Mcfadzien, Sandeep Rai, John Nordgren, Carl Bruch, Dan Schramm, and Cassandra Brooke. Their candid commentary improved the book greatly.

We thank Barbara Dean, Todd Baldwin, and Erin Johnson at Island Press for striking such a nice balance between encouragement and harassment, and Marce Rackstraw for being patient, expeditious, and responsive and for turning out such lovely graphics. We also wish to thank Carlos Drews, Nick Lunn, Dee Boersma, Wendy Foden, Mark Anderson, and Anton Seimon for letting us tell their stories, Bruce Stein for his wonderful sunburn analogy, Earl Saxon for collaboration on the creation of figure 3.2, and Chris Bergh for providing us with the Key deer maps for figure 8.1. Mar Wonham provided research help for chapter 12, Meade Krosby for chapter 10, and Tasha Belfiore for chapters 8 and 9. The Kresge Foundation provided the funding that gave us the time to write this book while still allowing us to pay our rent.

We are indebted to our friends, family, and coworkers who put up with us as we periodically turned into writing zombies. They kept the home fires burning (quite literally for JRH, since she and Daniel heat with wood), made sure that work got done, and were really quite pleasant about it all. Special thanks to Eric Mielbrecht for telling us this would take more time than we thought but not making too big a deal about it when it did.

We are also grateful to all the people we have interacted with over the years who have helped us to refine the ideas presented here. This includes everyone who attended a Climate Camp, talked to us on the phone, participated in a class we taught, or stopped us to talk in a hall outside a meeting.

This book would not have been possible without the great libraries in our lives: American University Jack and Dorothy G. Bender Library, the Seattle Public Library, Bainbridge Island Public Library of the Kitsap Regional Libraries, the University of California Library system, and the University of Washington Library system. We made every effort to reduce greenhouse gas emissions, so we would like to thank the good public transit systems on which we rely (Washington State Ferry System, King and Kitsap County Metro Transit systems, Washington Metropolitan Transit Authority), our bicycles (Lara's Blue Mariah and Jennie's beloved Bridgestone that was, alas, stolen near the end of the book-writing process), and good shoes for walking (left and right).

Last but not least, we are very, very grateful to everyone who picks up this book and decides to include the reality of climate change in some decisions they make.

# Chapter 1

## *In the Beginning*

> We stand now where two roads diverge. But unlike the roads in Robert Frost's familiar poem, they are not equally fair. The road we have long been traveling is deceptively easy, a smooth superhighway on which we progress with great speed, but at its end lies disaster. The other fork of the road—the one "less traveled by"—offers our last, our only chance to reach a destination that assures the preservation of the earth.
>
> —*Rachel Carson,* Silent Spring

We are at a crossroads—or perhaps a traffic circle—of options about our future, including decisions about how we react to the reality of climate change. We must decide not only what to do about greenhouse gas emissions but also how to respond to the myriad effects of climate change as they continue to manifest themselves around our planet. Included in these choices is how we rethink natural resource conservation and management in light of climate change.

For more than a century the collective focus has been on protecting resources as they are, restoring them to what they were at some previous time, or using them based on past experience and understanding. Unfortunately, past and even present conditions are not likely to resemble the future. We are already seeing alterations in the natural world as a result of climate change. Warmer temperatures, different precipitation patterns, rising sea levels, acidifying waters, and greater climatic variability are leading us to new and ever-changing environmental conditions. This means we need to reconsider our goals and objectives and the tools we use to meet them. We may not need to abandon our goals, but we certainly need to examine thoughtfully how to achieve them given this new reality.

**BOX 1.1 ADAPTATION**

**Unless otherwise noted, the term *adaptation* in this book refers to human efforts to reduce the negative effects of or respond to climate change, rather than evolutionary or biological adaptation.**

In this book we will explore how the world is changing and how our perspective can adjust to keep up when it comes to protecting and managing nature and the resources it provides. We will begin with an exploration of climate change basics and a look at where the world is today. Obviously this small tome cannot cover the myriad changes afoot, nor provide a detailed exposition on how climate change and adaptation may play out in every corner of the world. It certainly cannot tell you with certainty what the future will be—no one can. We hope, however, that it gives you a broad-brush outline to flesh out based on your own local knowledge. At the very least it may help you to avoid overlooking key categories of climate change impact and vulnerability that may be lying in wait to thwart your long-term success.

With climate change basics as a foundation, we explore what is meant by the term *adaptation* in the climate world. In any field, some terms get bandied about with no clear sense of what they mean, and *adaptation* runs the risk of being such a term. Soulé (1986) posited that the creation of the field of conservation biology was possible only when there was a critical mass of people who self-identified as conservation biologists. In the case of climate change adaptation (as distinct from evolutionary adaptation; see box 1.1), the term appeared in the 1992 United Nations Framework Convention on Climate Change, well before there was a critical mass of practitioners behind it. The number of climate adaptation practitioners is growing, but the field is still poorly defined and rapidly evolving. Because of the potential for adaptation-specific funding, some groups are working hard to define adaptation based on what would bring them funding rather than what would best reduce vulnerability to climate change. Even the terminology itself is confusing (see, for example, box 1.2). Policy and management decisions are moving ahead despite these limitations, so we must build a common understanding of what we are all working toward. This book attempts to lay out a framework, or philosophy, to help move this dialogue along.

Having a philosophy or a framework is all very well, but without techniques and tools you cannot get very far. We spend a good deal of the book exploring the wealth of conservation and resource management tools, their vulnerabilities to climate change, and how they can be implemented in ways that maintain or increase their effectiveness. This includes many old friends—protected areas; species-based protection; connectivity; regulating harvests; reduction of pollutants; control of invasive species, pests, and disease; and restoration—but adds a new spin to how they can be applied to deal with climate change.

**BOX 1.2 POINT OF CLARIFICATION: ADAPTIVE MANAGEMENT**

Adaptive management is not synonymous with climate-savvy management. It is a research approach to management in which practitioners consciously experiment with, learn from, and improve the efficacy of their actions. It has three elements: testing assumptions through monitoring and experiment, adapting assumptions and interventions based on new information, and learning. If the new information relates to climate change this could in fact lead to climate change adaptation, but adaptive management per se is not about climate change. Simply using adaptive management does not itself guarantee adaptation to climate change, although people sometimes mistakenly use the term this way. Climate change must be explicitly considered in all elements of an adaptive management plan for that plan to be truly climate savvy.

Along the way we offer some thoughts on how to use models, a mainstay of climate change science and planning, as well as options for integrating the needs of humans and nature to increase the likelihood of success for both and how to use and improve governance mechanisms to support adaptation efforts. And, of course, we explore the most important tool for developing and implementing adaptation: creative thinking.

## A Brief History of Adaptation

Following the completion of the Intergovernmental Panel on Climate Change's (IPCC) first Impacts Assessment in 1990, there was an identified need for a standard framework to create comparable data across studies. This led to the IPCC Technical

**BOX 1.3 ENVIRONMENTAL DEGRADATION AND ENERGY: FORESTS TO RIVERS TO FOSSIL FUELS**

Climate change is not the first adverse effect of the human need for energy. The use of wood to heat and cook is a sustainable activity when population densities are low, but as populations have grown deforestation has spread out around population centers. The flow of rivers has long been harnessed—from simple water wheels and mills to massive hydroelectric processes—in ways that changed flow regimes, river temperatures, and connectivity of waterways. Damage from extraction (mining, oil and gas drilling), transport (infrastructure and spills), and combustion (smog, acid rain) have all had dramatic effects on our planet. It seems we have yet to identify large-scale energy paradigms that are less at odds with our environmental conservation aspirations.

Guidelines for Assessing Climate Change Impacts and Adaptation (Carter et al. 1994). These early guidelines laid out some definitional and mechanistic needs that the inorganically derived field of adaptation required (see box 1.4). As mentioned, unlike most fields where a group of interested parties creates a discipline from the bottom up, adaptation has been created from the top down almost by edict from the IPCC and the United Nations Framework Convention on Climate Change. As a result, adaptation has no formal discipline to which to refer, no evolutionary or reverential literature, and no pedagogical process or best practices for training new practitioners. It is a field almost reinvented by each new participant. This is a challenge for a field that requires urgent translation of concepts into practice if it hopes to be effective.

Although climate change adaptation is still developing, it can take advantage of what we have learned about effective and ineffective planning and practice for resource management and conservation. Traditional and modern approaches to resource management can be blended into a more holistic framework upon which to base a new, climate-savvy approach. There is an opportunity for adaptation to reintegrate the human element, designing conservation for whole systems rather than trying to separate "pristine wilderness" and human communities. In a way, adaptation could allow for the correction of some of our past mistakes in conservation and management, just as its very reason for being is correcting some of our past mistakes in terms of poor energy planning.

One early list of adaptation approaches included eight categories: bear losses, share losses, modify the threat, prevent effects, change use, change location, research, and

### BOX 1.4 POINT OF CLARIFICATION: MITIGATION AND ADAPTATION

In the lexicon of the United Nations Framework Convention on Climate Change, climate change is addressed from two perspectives: mitigation and adaptation. Mitigation addresses the root cause of climate change, limiting emissions of greenhouse gases or increasing their removal from the atmosphere. Adaptation refers to human actions taken to limit the negative or take advantage of the positive effects of climate change. These responses can be proactive (anticipatory) as well as reactive.

Adaptation and mitigation are not choices to be weighed against each other: both are necessary responses to the challenge of climate change. Effective adaptation depends on effective mitigation to slow the rate and extent of climate change to an adaptable amount. Climate change has already progressed to the point where some effects are unavoidable, making adaptation likewise unavoidable. Even mitigation efforts will need to adapt, as many lower-emissions energy sources such as nuclear (which relies on water for cooling) or hydropower are vulnerable to climate change. It is incumbent on individuals working on both the mitigation and adaptation components of the climate change issue to not only understand but also support the activities of their counterparts.

educate (Burton et al. 1998). In the years since, other frameworks have appeared that parse adaptation options in other ways. Such lists are useful, but they are only broad, general starting points. Our aim in this book is to provide a bit more detail and a few of the lessons that have been learned in the decade since that list was devised.

Adaptation is more than a simple list of options, or even a complicated list of options. It is a complex, ongoing process and a state of mind. If adaptation were simple, we could tell you what to do and you would be guaranteed success every time. When was the last time a *recipe* worked for resource management even without climate change? With uncertainty about future climate trajectories, the biological responses to those climate changes, and the human responses to both, not even a really smart flowchart can plot the course. If we instead think of climate adaptation as a set of informed actions we take based on an awareness of climate change and associated uncertainties, then we have a lifestyle rather than a life sentence from which to work.

## You Do Not Have to Reinvent the Wheel

Climate change may present a whole new challenge, but it also offers an opportunity to make conservation and resource management more robust as we shift from recovering from past damage to preparing for future changes. We can build this new path standing on the shoulders of giants as we take advantage of all we have already learned about how to practice conservation biology and resource management.

Conservation biology, resource management, ecological restoration, and climate change adaptation can all be viewed as crisis disciplines in that they require us to act with incomplete knowledge, a tolerance of uncertainty, and a mix of science, art, intuition, and information (Soulé 1985). Yet while these elements have been part of conservation, management, and restoration from the get-go, they are somehow paralyzing for practitioners when it comes to climate adaptation. Many of the leaps of faith required for climate change adaptation are the same leaps we have been taking as part of our daily practice for years. In fact, once you start thinking about how to incorporate climate change into conservation or management it will seem curiously similar, or at least analogous, to what is already standard practice.

## Final Thoughts

We are all on a mission to protect ecosystems, support sustainable development, manage natural resources for ongoing use, and protect human well-being. Succeeding at this mission now requires the inclusion of climate change in our philosophy and practice. Many of our old tools, and the ways in which we use them, need to be modified if we are to meet the goal of our mission. Unfortunately, time is of the essence because climatic changes are happening now and, without our intervention, will continue well into the future.

# Chapter 2

# *Climate Change and Its Effects*

## WHAT YOU NEED TO KNOW

To see what is in front of one's nose needs a constant struggle.
—*George Orwell*

For better or worse, climate change is affecting many elements of the world around us. We can incorporate this reality into our planning or we can ignore it, but the climatic changes currently under way will continue regardless. Species ranges will continue to shift, the timing of seasonal events will continue to change, and weather patterns will no longer follow familiar paths. If we fail to look at how our policies and practices might be affected by these changes, we run the risk of investing time, money, and political capital in plans that are at best irrelevant and at worst maladaptive. This is true for any sector or activity influenced by climatic conditions, be it resource management, development, or conservation. Climate change is not the only important consideration for conservation or natural-resource planning, but ignoring it would be as shortsighted as ignoring the possible influence of land use, pollution, or invasive species.

To adapt conservation to inevitable climatic changes, practitioners need a solid understanding of the basics of climate and climate change: what we know, how we know it, and how certain we are. Both the range of plausible changes and the degree of uncertainty are central elements of how and why "business as usual" conservation is no longer an option (or at least not an option with a high likelihood of success). The range of possibilities matters because it highlights vulnerabilities of current approaches, such as basing conservation plans only on the current distribution of species whose range is already shifting. The uncertainty matters because even the best climate models cannot provide 100 percent certainty about the future climate, so we need to focus on how best to plan in the face of uncertainty rather than just on improving the models.

The goal of this chapter is to provide an overview of the sorts of physical, chemical, and biological changes the future may have in store. The Intergovernmental Panel on Climate Change's fourth assessment report (IPCC 2007a, 2007b) is an excellent source for more details on the material presented here, but neither this chapter nor the IPCC is any substitute for a place-based assessment of current climatology, future changes, their effects, and potential vulnerabilities.

## Climate Variability versus Climate Change

Day to day, season to season, year to year, and decade to decade, species and ecosystems experience changing climatic conditions. In contrast, directional climate change is a longer-term trend toward a new climate regime. The current rate of warming is roughly ten times faster than any rate in the last 10,000 years (IPCC 2007a). As illustrated in figure 2.1, current atmospheric concentrations of carbon dioxide, a major driver of climate change, are significantly higher than at any time in the last 400,000 years. After millennia of cycling between roughly 180 and 300 parts per million, carbon dioxide levels jumped from 280 to 385 in just over a century, and at current rates of emission are likely to reach an unprecedented 800 parts per million within the next century. The

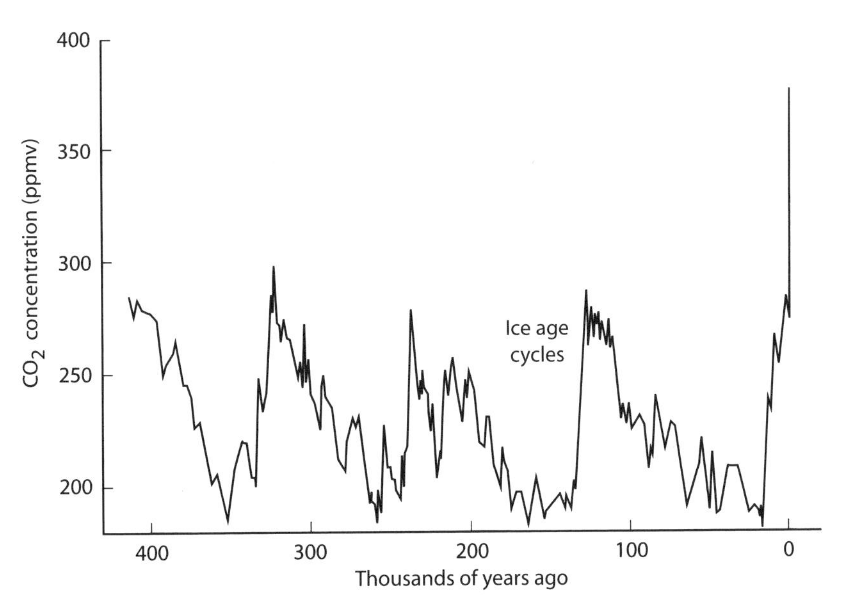


FIGURE 2.1 Earth's atmospheric carbon dioxide concentration over the last 400,000 years as measured from ice cores and air samples. Data from Neftel et al. 1994; Petit et al. 1999; Keeling and Whorf 2004; Monnin et al. 2004.

past 10,000 years have been remarkably stable climatically, and the species, communities, and cultures existing today have evolved in the context of this stability.

Climatic changes occur on a number of spatial scales. Global changes such as we are currently experiencing happen only when there is some shift in the forces that determine Earth's total heat balance, such as the position of Earth relative to the sun or the concentration of heat-trapping gases in the atmosphere. Regional climate change may result from global change, but it may also result from a redistribution of heat without large changes in global mean temperature, as happens when there are significant changes in major ocean currents. An example of extreme regional changes not linked to global change are several temperature increases of between 10 and 16°C during the course of just a few decades in Greenland during the last glacial period (IPCC 2007a). A less abrupt but global change was Earth's transition from the last ice age into the current Holocene Period roughly 10,000 years ago. Following a prolonged period of generally cool but highly variable conditions, the global average temperature increased on the order of 5 to 7°C, and sea level rose by 100 meters as ice caps and glaciers melted and oceans warmed and expanded.

Annual or decadal changes occur on top of the current directional global warming. The El Niño–Southern Oscillation, or ENSO, can bring large changes in sea level, temperature, and precipitation on a year-to-year basis. The Pacific Decadal Oscillation (PDO), in contrast, switches between warm-wet and cool-dry phases over the course of decades (fig. 2.2). Similar decadal oscillations exist in other regions of the globe. Thus the recent extreme warmth in parts of the world is due to the combination of the warm-phase PDO and directional global warming. As the PDO shifts to a cool phase, we should expect a period of cooler weather. This temporary cooling in no way indicates an end to global warming, however, and when the PDO shifts back to a warm phase, we can expect it to be even warmer than during the previous warm phase.

Another way to put this is that directional climate change does not mean that each year will be warmer than the last. It means that on average, over scales much longer than decades, the Earth's atmosphere is warming. There will be warm years and cool years, warm decades and cool decades, but on average, over the next 100 years things will keep getting warmer.

## Measuring Change, Predicting the Future

Scientific knowledge about past climates and climatic changes comes from a variety of sources. Air bubbles trapped in ice provide information on the concentration of greenhouse gases in the atmosphere when the air bubbles were formed. The ratio of different isotopes of oxygen or nitrogen in the ice itself indicates how much of Earth's freshwater was in solid versus liquid forms (isotopes are different versions of the same element that have different weights and therefore are more or less likely to end up in the atmosphere as the globe heats or cools). The distribution of fossil pollen or marine microorganisms with varying temperature tolerances provides yet another indicator for

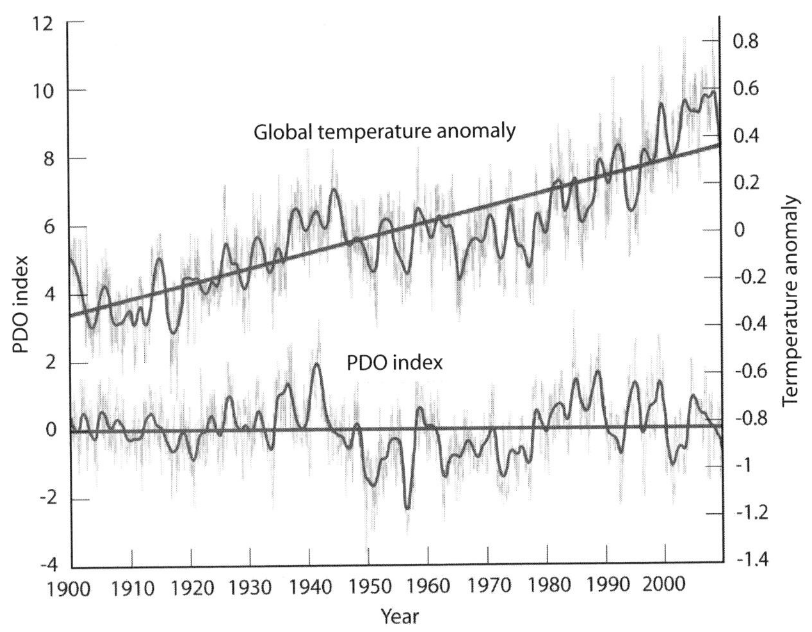


FIGURE 2.2 Comparison of the Pacific Decadal Oscillation (PDO) with overall global temperature trends. The upper line shows average annual global temperatures for the last century, expressed relative to the 1961 to 1990 average. The lower line shows the PDO Index, a measure of decadal-scale climate patterns. Values greater than zero indicate warm-phase PDO, while negative values indicate cool-phase PDO. In the 1930s and 1940s, a warm-phase PDO combined with the overall increase in global temperature to create a period of rapid warming. From roughly 1946 through 1977, a cool-phase PDO canceled out the background warming trend, followed by another period of rapid warming when the PDO switched back to a warm phase. Data from the National Oceanic and Atmospheric Administration Global Land-Ocean Temperature Index (temperature anomaly line) and Climate Impacts Group/Nate Mantua (PDO).

past climatic conditions. Combining information from a number of these "climate proxies" builds a more complete picture of past climate conditions and helps determine how certain we can be about our conclusions.

In combination with these measurements of past conditions, scientists develop climate models to understand past climates and to project into the future. Models are based on first principles—that is, on our understanding of the physical, chemical, and biological processes that govern the climate system. These includes such factors as atmospheric concentrations of heat-trapping or heat-reflecting gases, how much heat Earth's surface absorbs or reflects, ocean currents, and plant cover. The accuracy of models is verified by testing their ability to simulate current or past climates. The better a model simulates these known climates, the more weight we give its projections of future climate. Scientists also compare the results of multiple models as a way of better estimating degrees of certainty. If all models agree on certain projections for the future, we can have greater confidence in those projections.

**BOX 2.1 ARE HUMANS TO BLAME?**

**From an adaptation perspective, this isn't the central question. Adaptation focuses on what we can do to reduce vulnerability to climate change. Will reducing greenhouse gas emissions reduce the rate and extent of climate change? Almost certainly, even if a significant amount of that change is due to natural causes. The geologic record suggests that $CO_2$ is very likely to have amplified even those past warming periods for which it may not have been the trigger.**

Although climate change is a global phenomenon, conservation and management typically take place locally or regionally. Thus the output from global climate models with a spatial resolution on the order of hundreds of kilometers is not ideal for management planning. Climatologists have developed a number of approaches for generating climate projections on finer scales. The two most commonly used approaches are statistical downscaling, where statistical methods are used to relate large-scale model outputs to local surface variables such as topography or large bodies of water, and dynamical downscaling, where high-resolution regional climate models are embedded in global models. Different climate modeling approaches and their strengths and weaknesses are covered in more detail in chapter 7. Regardless of what type is used, climate models are only as good as the data you put into them, and increased spatial resolution does *not* translate into increased certainty.

Before leaping into downscaled climate models for a particular region, it is essential to consider the availability, quality, and relevance of existing climate data. These data are what is used to test the accuracy of downscaled models, and the quality and relevance of the models are only as good as the data used to create and verify them. In Madagascar, for instance, the temperature and precipitation trends measured by official meteorological stations, which are primarily in heavily disturbed areas, do not generally parallel the changes measured by researchers' stations inside forests. Without incorporating data from within forests as well as from without, there will be no way to know whether the regional climate models are relevant for forested areas. Investing time and money in downscaled modeling should be done only after developing an understanding of regional climatology as it functions today, determining whether increased spatial resolution will be useful, and acknowledging the degree of uncertainty downscaled models will have for your region.

## Physiochemical Changes

Given the uncertainties inherent in predicting the future, what management-relevant changes might we expect in the next few decades? We present here a summary of current projections for a number of physical and chemical variables, but practitioners should remember that while there are clear global trends, there is also significant local and regional variability.

### *Temperature*

The average global temperature is rising and will continue to do so. The rate of change varies across the globe, with the Arctic and much of the Antarctic Peninsula experiencing the most rapid changes. Rapid Arctic warming is due at least in part to the positive feedback generated by melting snow and ice: melting snow and ice exposes the darker earth and water that lie underneath, more heat is absorbed rather than reflected, and the rate of warming increases. On a local or regional level, temperature changes may be significantly influenced by land use and vegetation changes. Deforestation generally increases local warming, as does increasing the amount of bare earth or pavement. Annual or decadal cycles such as the North Atlantic Oscillation will be overlaid on the long-term warming trend, leading to warmer or cooler years or decades. Thus there may be years or decades of more benign conditions followed by years or decades in which conditions become particularly stressful (fig. 2.2).

### *Precipitation and Storms*

The amount, timing, and type of precipitation are all changing. In many areas, both drought and flooding are predicted to become more extreme, a change that will be exacerbated in areas where river flow is heavily dependent on snowpack. For instance, the western United States has seen a trend toward more rain and less snow during the last half century (Knowles et al. 2006), and the peak spring flooding is occurring earlier in part due to earlier snowmelt. The shift toward rain over snow and earlier snowmelt also contributes to lower river and lake levels in summer. Reduced dry-season precipitation combined with increased release of water by plants and evaporation due to higher temperatures will further worsen drought.

Land use changes contribute both to changes in precipitation and to the effects of those changes. Loss of forest cover increases water loss and reduces the amount of moisture in the air in the surrounding region, leading to drier conditions. Loss of vegetation on hillsides worsens flooding and erosion during heavy rain, and loss of wetlands likewise worsens flooding by reducing the ability of ecosystems to absorb floodwater.

The science surrounding the effect of climate change on hurricane or cyclone frequency and intensity is not yet clear, and the nature of historical trends is controversial due to the patchy nature of storm detection prior to the advent of global satellite coverage in the 1970s. From a theoretical standpoint, the expected effect of climate change on storm frequency and intensity depends on the relative importance of various elements of the climate system, such as absolute sea surface temperature, relative sea surface temperature across ocean basins, lower stratospheric temperature, and wind shear. If absolute sea surface temperature alone is the primary determinant of storm frequency and intensity, we can expect an unprecedented increase in storminess. If wind shear is more important, some regions would see a decline in storm intensity and frequency. Recent models incorporating a variety of factors and climate models suggest an overall trend toward fewer but more intense storms, with more rain carried by each storm.

Regardless of long-term trends in storm frequency and intensity, sea-level rise will increase flooding and erosion risk along coasts. The increase in heavy development along the world's coastlines as well as increasing deforestation along vulnerable hillsides greatly increases the vulnerability of both human and natural communities to large storm events.

### *Fire*

Increased temperatures and drought are leading to increasing frequency and intensity of wildfires in many regions, even those with minimal human activity (Westerling et al. 2006). Higher atmospheric carbon dioxide may lead to greater plant growth, providing more biomass for fires. Drought stress may mean more dead plant matter to provide easy tinder in the short term, but over the long term it may reduce fire risk by reducing the amount of plant matter available to fuel fires. This is the scenario predicted for Southern California, where increasing drought is projected to shift much currently forested land to shrubs and grasses instead (Westerling et al. 2006). Increasing frequency and intensity of insect outbreaks may likewise create large swaths of dead trees to fuel fires in the near term, followed by a period of lower fire risk once the dead trees have burned or decayed. Increased fire frequency is also likely to cause a shift toward more fire-resistant or less fire-prone plant types, ultimately lowering fire risk but leading to dramatic changes in ecosystem types. In the Amazon, the increase in natural fires combined with ongoing climate change and deforestation for agriculture or other uses may lead to the loss of most of the existing rainforest within the century (Huntingford et al. 2008).

Increasing fire frequency and intensity is also likely to speed the rate of climate change both locally and globally. A large, hot fire may release in hours carbon that it has taken decades for a forest to store, instantly increasing the amount of carbon dioxide in the atmosphere and decreasing the ecosystem's capacity to take up new carbon. Ash from fires upwind from coastal areas may stimulate the growth of some marine organisms. The loss of forest cover caused by large fires also increases temperatures and dryness regionally.

### *Snow, Ice, and Frozen Ground*

Globally, more glacier, ice cap, and ice sheet area is being lost each year than is being gained. The rate of loss varies from year to year and there are some areas where glaciers or ice sheets are growing, but this does not negate the global trend. For instance, several glaciers in Greenland experienced two years of very rapid shrinking, but then returned to "normal" levels. Glaciers in the Western Himalayas are growing due to a regionalized decrease in summer temperatures and an increase in precipitation, while glaciers in the rest of the Himalayas, which supply more than half of the water for the Ganges and other major Asian rivers in summer, are shrinking (Fowler and Archer 2006). Taken by climate change deniers as evidence that there is no global warming, these examples instead illustrate the complex nature of how global change plays out at the regional or local level. Some areas with glacier-dependent water sources are already seeing a drop in summer water supply, and continued loss of glaciers will lead to loss of freshwater supplies for cities and ecosystems. Planning now for future decreases in water availability will lessen the negative effects of those decreases.

Melting glaciers increase the size and number of glacial lakes, increasing the risk of glacial outburst floods as lakes of meltwater build up behind temporary dams of ice or debris. Such a flood destroyed one of Tibet's major barley-producing areas in 2000, taking out at least 10,000 homes and almost 100 bridges and dikes. Nepal alone has twenty-seven glacial lakes with the potential to cause catastrophic flooding.

In the Arctic, loss of sea ice appears to be accelerating, with some experts predicting ice-free summers by 2030 or even earlier (Wang and Overland 2009). Sea ice is also declining in the Southern Ocean, although not as rapidly. Loss of sea ice may increase in-water primary productivity by allowing more light to reach the water, but it will decrease rich areas of productivity generally associated with the ice edge. Because sea ice reduces wave intensity, loss of coastal sea ice is also leading to dramatic increases in coastal erosion rates around the Arctic.

Melting permafrost and ground ice is another problem for polar ecosystems. In addition to increasing ground instability, melting permafrost changes lake dynamics. During the early stages of melting, lakes may be created or expand as a result of buckling ground and the melting itself. As melting progresses, the layer of permafrost that prevents the water from draining away may be lost, at which point the lake drains completely. Between 1973 and 1998, one region of Siberia lost more than 1,000 large lakes due to melting permafrost (Smith et al. 2005). Melting permafrost also releases methane, a greenhouse gas with a warming potential roughly twenty-one times as powerful as that of carbon dioxide.

### *Freshwater Ecosystems*

Climate change is affecting the physical and chemical nature of freshwater systems in three interconnected ways: it is altering water quality (e.g., temperature, pH, or clarity), water quantity, and water timing (e.g., seasonal flooding peaks).

As mentioned previously, earlier snowmelt is leading to earlier peaks in spring flooding. The shift from snow to rain in winter has led to more winter floods or higher winter water levels and lower summer water levels in many areas, altering the availability of permanent lake and pond habitat at high altitude. The predicted drop in snowpack in the western United States (as much as 60 percent; Leung et al. 2004), for instance, means that many year-round streams and lakes may become seasonal.

Rivers, lakes, and streams in many regions are warming as a combined result of increased air temperature, lower water levels, and slower flow. As warming increases the temperature (and therefore density) difference between surface and deep water, the seasonal overturning that mixes nutrients from deeper water back into the surface water may weaken or stop, decreasing productivity in surface waters. If nutrients are not limiting, however, increased water temperature can trigger dramatic algal blooms, which may lead to low-oxygen zones as the algae and other organic matter decay. Contaminant loads in streams and lakes may experience periodic increases as drier conditions reduce water volume, as larger floods mobilize more soil-bound contaminants, and as longer dry periods allow higher concentrations of contaminants to build up on land prior to being washed into waterways during the first rains.

### *Marine and Coastal Regions*

Sea-level rise is one of the most straightforward effects of climate change. Globally, sea level is rising primarily due to the thermal expansion of warming water and water input from melting ice caps and glaciers. As with other elements of global change, how fast and how much sea level rises will vary over both space and time. The movement of geologic plates can warp Earth's crust, dramatically altering rates of sea-level change. On the northwest corner of Washington's Olympic Peninsula, such tectonic forces are causing land to rise, and relative local sea level is consequently dropping. Just a few hundred kilometers away in the South Puget Sound, the same tectonic forces are causing land to sink, and relative local sea level is rising much faster than the global average. Ice-cap melting will eventually outpace tectonic uplift on the Olympic Peninsula, but sea-level rise there will remain lower than elsewhere (fig. 2.3). Deformation of the crust also occurs in response to the addition or release of weight, such as when glaciers expand or retreat, large lakes are created or drained, or sediment load changes. Areas that were covered in ice during the last ice age are still springing back (in a slow, geological way) from the weight of ice that was lifted from them, causing a local drop in sea level. In many large deltas, the weight of accumulated sediments carried by rivers causes the land to sink; it also sinks as a result of sediments compressing over time. If the rate at which new sediment is delivered keeps pace, there may be no net effect. In most deltas, however, sea level is rising faster than the global average.

The oceans have absorbed roughly 40 percent of the carbon dioxide emitted since the start of the Industrial Revolution. Through simple, well-understood chemical processes, absorbing carbon dioxide decreases the pH of ocean water, and the average pH of the oceans has dropped by 0.1 pH units since 1750. Tropical oceans are changing more slowly than high-latitude ones. Decreasing pH has caused the calcium carbonate saturation horizon, or the depth below which calcium carbonate dissolves, to rise by up to 200 meters. Undersaturated water (that is, water in which calcium carbonate dissolves) has already appeared along some coasts during strong upwelling events (Feely et al. 2008), and surface waters in the Southern Ocean and subarctic Pacific Ocean will become corrosive to calcium carbonate within the next century.

Ocean salinity is also changing. In areas with high evaporation rates and no increase in precipitation, the salinity of the surface layer has increased measurably. Such areas include most of the Indian Ocean and the tropical and subtropical Atlantic. In contrast, much of the Pacific and subpolar oceans in both hemispheres is becoming less

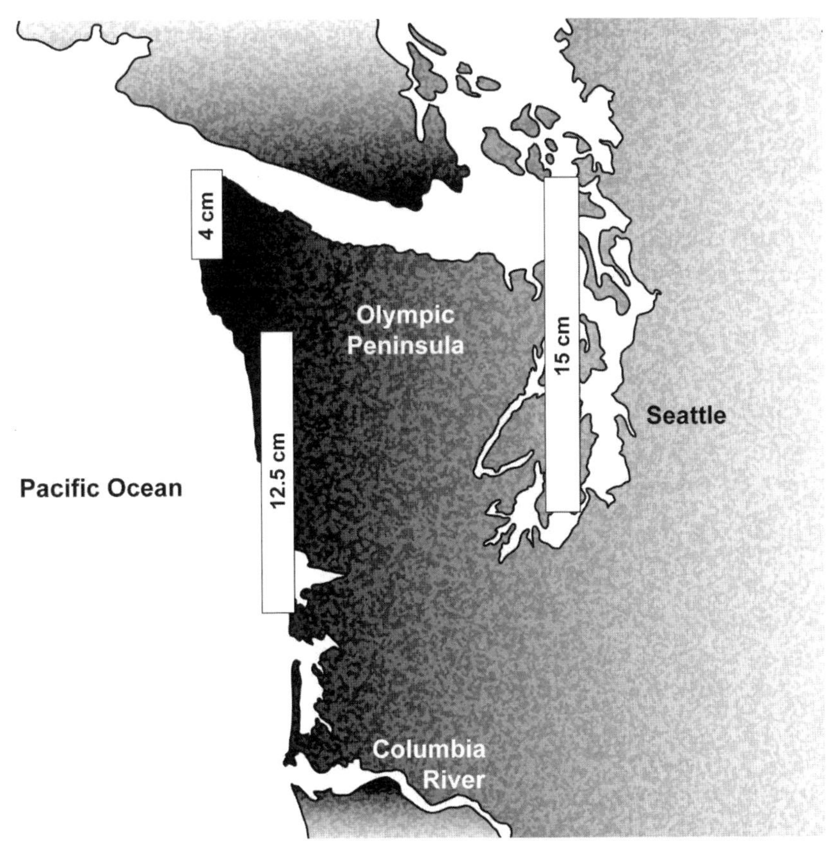


FIGURE 2.3 Projected sea-level rise by 2050 for three locations in Washington State. Values from mid-range sea-level rise models in Mote et al. 2008.

salty, primarily because of changes in precipitation. In some areas, melting sea ice and increasing freshwater input from land also play a role. These trends in salinity are expected to continue.

One final physiochemical change in Earth's oceans is decreasing oxygen concentrations in some layers of the ocean, and an expansion of this so-called mid-water low-oxygen zone into both deeper and shallower waters. Because warmer water holds less oxygen, decreases in marine oxygen concentrations are likely to continue well into the future, with metabolic consequences for many marine animals. Imagine how much more difficult physical activity would become in a world with less oxygen!

## Biological Changes

The physical and chemical changes described in the previous section will lead to potentially profound biological changes on the level of individuals, populations, communities, and ecosystems. These changes may be transient and reversible, or they may be long-term and irreversible, such as species extinction. Some changes reflect physiological, behavioral, or evolutionary adaptation (that is, changes in individuals or populations that decrease the negative effects of change), while others are manifestations of increased stress.

Climate change influences individual organisms in a variety of ways including metabolic rate, growth rate, time to maturity, and overall health. On the population level, these individual effects translate into changes in key demographic variables such as birth and death rates, and may alter a population's evolutionary trajectory by changing which features are favored by natural or sexual selection. For species-based conservation, understanding how climate change will affect these basic elements of population dynamics is essential.

The geologic record, including pollen deposits and fossils of organisms ranging from trees to microscopic sea creatures, offers a sense of how species and communities may respond to climate change. Ecosystem types shift, expand, or disappear, species ranges change, and some species flourish while others go extinct. Although the reigning view was once that entire ecological communities shift together, tracking appropriate climatic conditions, it appears that different species shift at different rates. Thus plant communities during the transition between warm and cold periods look nothing like either the communities that precede them or those that follow.

Assessing the vulnerability of organisms and ecosystems to climate change is not a matter of simply predicting the magnitude of change. Populations and communities that have evolved in areas with less climatic variability, such as the tropics, may be more sensitive to change than those that see huge differences in temperature and precipitation on a daily or annual basis (Tewksbury et al. 2008). Population-level responses also depend on changes in food availability, competition, and predation and, for some organisms, on soil type or day length. The ecological implications of climate change can be explored from a broad-scale geographic point of view or from a bottom-up look at effects on individuals, populations, and communities. The ongoing interplay between these top-down and bottom-up approaches has greatly enriched our understanding of how the world works.

### *Parasites, Predators, and Competition*

Although models of current and future species distribution commonly address only climatic variables, a species' actual range is determined by the interplay between these variables and a range of biological factors. Climate change will affect species distribution not just through physical and chemical changes, but by changing the interactions among species or through differential effects on interacting species. A species may be excluded from appropriate habitat by a competitively superior species or by the absence of a key food item just as easily as by climatic factors. For instance, some species of algae that are negatively affected by ultraviolet radiation still do better in areas exposed to UV because the animals that eat them are more sensitive to UV than the algae (Bothwell et al. 1994).

The effect of climate change on pest and parasite populations is obvious to anyone driving through areas of North America where various species of bark beetle have laid waste to millions of acres of formerly evergreen forest. Higher average temperatures mean beetles mature faster and have more generations per year, and less extreme winter temperatures mean beetle populations are less often killed off by hard freezes. While bark-beetle die-offs are a natural and important element of mountain conifer ecosystems, the increased rate and extent of die-offs is new and may reshape ecosystem function and diversity. In particular, the ability of lodgepole pine forests to regenerate following a die-off depends on the presence of adequate numbers of seeds in the soil. In areas where most lodgepoles are less than 100 years old, there may not be enough seeds to support regeneration (Sibold et al. 2007).

Other examples of warming's effects on parasite ranges or infection dynamics include the oyster diseases dermo and MSX along the east coast of the United States (Ford and Smolowitz 2007) and muskox lungworm in the Arctic (Kutz et al. 2004).

### *Changes in Timing*

There have already been widespread shifts in the phenology, or timing, of certain seasonal events. In many cases, spring events such as bud-burst and nesting are occurring earlier, and fall events such as leaf-loss of deciduous trees are occurring later (Parmesan 2006). Some migratory birds are arriving earlier to their breeding grounds and later to their wintering ground. In and of themselves such shifts may be adaptive. Birds that have their first clutch of eggs early may be able to hatch out more than one clutch in a single season. The extended growing season at higher latitudes allows higher ecosystem productivity. Unless all species within a community shift timing in parallel, however, there will be some degree of phenological mismatch. Plants may bloom before their pollinators arrive, or animals may no longer have their young at the time of peak food availability. Such mismatch may be beneficial to one of the partners, such as when a plant flowers before bud-eating beetles arrive, or caterpillars pupate before predatory birds arrive, but the net effect will be a shift in the community structure. Species with particularly specialized relationships, such as those that rely on a single pollinator for fertilization, are most vulnerable to phenological shifts, while generalist species are more likely to successfully adjust.

### *Range Shifts*

Climate plays a large role in determining species ranges, so it is no surprise that many species have already shifted or expanded their ranges toward higher elevation or latitude as warming progresses. The Arbor Day Foundation has even changed its hardiness zone maps indicating which plants are likely to do well in different parts of the United States (fig. 2.4). Large-scale range shifts have been widely observed in the fossil record in response to previous periods of change. Some species expand their range, moving into new areas while remaining in old ones. Other species experience a range shift,

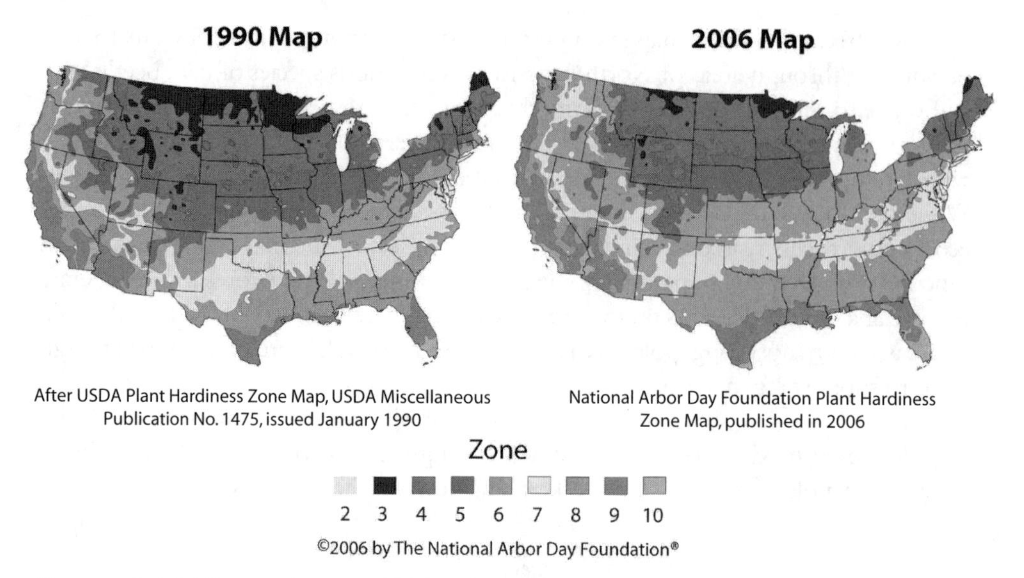


FIGURE 2.4 Differences between 1990 and 2006 hardiness zones for the United States. Hardiness zone is calculated using average annual minimum temperature, and is a good predictor for which plant species will do well in any given location. Maps based on those at www.arborday.org/media/map_change.cfm.

expanding into new areas while being forced out of old areas because climate change has rendered them unsuitable. A third group of species may see their range contract. This is happening to many montane and polar species, as well as those that live at continental margins, since they cannot go higher than the top of the mountains, or farther north or south than the poles. Terrestrial species on isolated islands may be doubly challenged: both sea-level rise and climatic changes are reducing the amount of suitable habitat, and range shifts across open water may not be possible.

Climate does not act alone to determine species' ranges. Some species require particular types of soil, for instance, and would be able to track suitable climate conditions only so long as there was suitable soil in which to grow. Interactions with other individuals or species are also a strong determinant of species distribution. As species expand into new areas, they may outcompete or prey on existing species, or they may come to coexist. The expansion of diseases and pests in particular can render an area unsuitable for species that previously lived there.

## Changes in Human Behavior and Resource Use

Patterns of human settlement are generally tied to natural resources such as the availability of freshwater, arable land, or abundant fish and game. Climate change is already altering the availability and reliability of these resources, and consequently is affecting where humans live and how they use natural resources. In Madagascar, for instance, some farmers are abandoning areas where farming is no longer feasible because of increasing drought and moving to the coast to take up fishing. Intensifying summer drought in the western United States has likewise amplified the ongoing tug-of-war for water among farmers, salmon advocates, and municipalities. Increasing rates of coastal erosion are leading some communities to install dikes, breakwaters, or other forms of shoreline hardening that interfere with the natural landward shift of coastal ecosystems in response to sea-level rise, as well as reducing existing habitat. All these changes have implications for conservation and resource management, and even organizations and individuals focused primarily on conservation rather than human welfare must take account of how humans will alter their behavior and resource use in the face of climate change. In some cases, development, disaster relief, and conservation organizations can find true win-win-win situations. In Vietnam, the Red Cross/Red Crescent Society initiated widespread mangrove restoration and replanting to reduce the rate of dike erosion; these efforts have also led to increased biodiversity, food security for the communities, and additional protection from coastal storm surge and flooding.

## Synergistic Effects

Climate change is not occurring in a vacuum. All the other problems that resource managers have been addressing for decades will continue to influence communities and ecosystems. In many cases, effects of climate change will interact with these other stressors in synergistic ways, worsening the effects of both.

### *Contaminants*

Climate change affects both the exposure to and effects of chemical contaminants on organisms and ecosystems around the globe. Sea-level rise can introduce toxic chemicals into near-shore environments if agricultural land, landfills, or other sources of toxic materials are flooded. Increasing storm intensity increases the risk of sudden flooding and release of toxic materials—as happened during Hurricane Katrina in New Orleans—and heavier rainfall may increase the release of contaminants from soil. Climate change is predicted to increase the seasonal influx of phosphorus into the Great Lakes, for instance, increasing the risk of algal blooms and dead zones. Even in areas where nutrient input remains constant, climate change is likely to increase the risk of harmful algal blooms and dead zones by increasing water temperature.

Temperature, pH, and salinity, all predicted to change as a result of climate change, affect the toxicity of various contaminants. Temperature may modify pollutant chemistry in ways that affect toxicity, and increasing temperature generally increases the uptake of pollutants via increasing metabolic rates. The uptake of many metals increases as salinity falls, so coastal waters subject to increasing freshwater runoff may see more problems with metal pollution even with no change in the level of metal pollution itself. This may be particularly problematic in enclosed or semi-enclosed seas such as the

Baltic Sea that are likely to see strong seasonal decreases in salinity as a result of heavy rains. On the flip side, contaminants may change the tolerance of organisms for variation in temperature, pH, or salinity. The highest survivable temperature for many freshwater fish species as well as for some invertebrates such as crabs or clams is decreased by some organic pollutants.

### *Land Use Change*

Deforestation provides the most dramatic examples of interactions between land use and climate change. Tropical deforestation, particularly through burning, is a significant contributor to global climate change, and an even more significant contributor to local or regional climate change. In areas like the Amazon or sub-Saharan Africa, the trend toward decreasing rainfall as a result of global climate change is exacerbated by decreasing rainfall as a result of deforestation. Together, global warming and loss of tropical lowland forest have created drier conditions in tropical montane cloud forests, leading to steep declines in frog and toad populations, upslope movement of many bird species, and species extinction even in protected areas such as Costa Rica's Monte Verde Cloud Forest Reserve.

Deforestation increases the vulnerability of people and ecosystems to other effects of climate change as well. In areas where episodes of intense rainfall are likely to increase, deforestation increases the risk of flooding, mudslides, and erosion. Over the long term, this could deplete the topsoil to the extent that recovery becomes difficult or impossible without massive intervention. Loss of riparian vegetation exacerbates increased water temperature in streams and lakes, further stressing freshwater ecosystems.

Deforestation and other types of habitat fragmentation also create barriers to some natural responses to climatic change. During past periods of dramatic climate change, there were far fewer barriers obstructing species' ability to track climatic changes. Now roads, cities, and agricultural land may all interfere with natural range shifts. They may also prevent genetic mixing that may aid the evolutionary adaptation of species to new conditions. Overharvest can also decrease the likelihood of evolutionary adaptation by reducing available genetic variability.

## Tipping Points

The geological record indicates that regional or global climate may make abrupt shifts from one state to another over relatively short periods of time when some critical threshold is crossed. In other words, there are "tipping points" in atmospheric greenhouse gas concentration beyond which some element of Earth's climatic or biotic system may change in ways that cannot be undone on human timescales. Past examples of abrupt climate change include the decades-long droughts contributing to the fall of civilizations including the Tang dynasty in China, classical Mayan civilization in Mesoamerica, the Akkadian Empire in Mesopotamia, and the medieval Khmer Kingdom in Southeast Asia.

**BOX 2.2 WHAT IS "DANGEROUS ANTHROPOGENIC INTERFERENCE IN THE CLIMATE SYSTEM"?**

**The degree of climate change that is considered "dangerous" depends on cultural values, location, and which tipping point you are concerned about. The IPCC's 2007 report states that a 1 to 3°C temperature increase above preindustrial levels commits us to massive loss of biodiversity, loss of the Greenland Ice Sheet and a consequent 7-meter rise in sea level, and a significant loss of volume in the glaciers that feed most of Asia's major river systems, all changes that many people place in the dangerous category.**

The possibility of such dramatic changes is less tractable from a planning perspective than the progressive changes currently under way, but should be kept in mind. Our best chance at stabilizing the climate and avoiding such shifts in climate regime is to limit those contributing factors over which we have control, such as greenhouse gas emissions and land use change. The possibility of unanticipated or massive changes in climatic and biotic systems underlines the need for a combination of specific adaptations to known threats and general resilience-building for both natural and human communities.

### *Ocean Circulation*

Ocean currents transfer huge amounts of heat from one part of the globe to another, and from the ocean to the atmosphere, heavily influencing regional and global climate. For instance, the North Atlantic meridional overturning circulation, of which the Gulf Stream is a part (fig. 2.5), carries warm surface water from the equator to the northern Atlantic while returning cooler, deeper water toward the equator, and is largely responsible for Europe's relatively mild climate. This current and many others are driven to a large extent by differences in water density: the tropical water cools as it heads toward the pole, gradually becoming denser and eventually sinking. Decreased salinity in Arctic surface water due to increased freshwater input from land or melting sea ice could potentially alter the density differential to the point where the overturning circulation slows or stops. While some slowing is likely over the next century, a complete shutdown is not.

### *Loss of Arctic Sea Ice and Collapse of the Greenland Ice Sheet*

Arctic sea ice volume has been declining for decades and experienced a sudden and dramatic drop in the summer of 2007. This sudden loss likely reflected decreased ice stability following decades of thinning combined with sustained windy conditions, but may signal the beginning of the end for Arctic sea ice. As ice disappears, it uncovers water (or land, in the case of ice sheets or glaciers) that absorbs more heat than did the

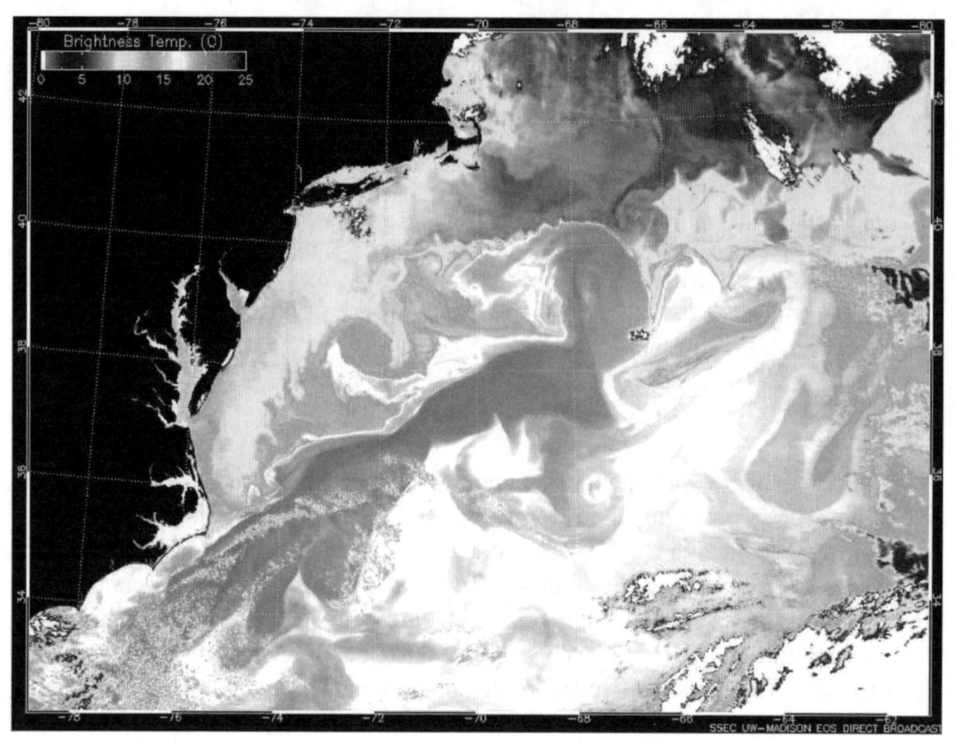


FIGURE 2.5 Satellite image of ocean surface temperature in the North Atlantic (North America in black on the left). The darker area stretching from bottom left to upper right represents the warm waters of the central Gulf Stream, with temperatures rapidly dropping off on either side. Original image from NASA's Earth Observatory.

ice, further accelerating local warming and ice loss. Once ice loss and local warming cross some threshold, this positive feedback loop makes complete loss difficult to halt. The Arctic Ocean will very likely be ice-free in summer by the middle of this century. This same process could play a role in the loss of the Greenland Ice Sheet. Once Earth reaches a threshold increase in temperature, somewhere in the range of 2 to 8°C, loss of the Greenland Ice Sheet will become irreversible, even if we are able to reduce greenhouse gas concentrations to levels that previously supported the continued existence of the Greenland Ice Sheet. The collapse of the Greenland Ice Sheet would raise global sea level by 7 meters over a century or two.

### *Loss of the Amazonian Rainforest*

Most climate models predict a significant drop in precipitation across the Amazonian basin. Increased drought stress would decrease vegetation cover, creating a positive feedback loop that would speed the drying process. Hotter, drier conditions increase the risk of wildfires, further threatening forest persistence. Lack of vegetation accelerates the loss of the thin tropical topsoil, making reestablishment of the original forest type more difficult. Even accounting for several levels of uncertainty, loss of much of the Amazonian rainforest by the end of the twenty-first century is likely (Huntingford et al. 2008).

### *Loss of Reef-Building Corals*

Climate change threatens reef-building corals by decreasing ocean pH, which makes it more difficult for corals to build their calcium carbonate skeletons, and by increasing water temperature, which increases the frequency and severity of bleaching events that damage or even kill corals. Bleaching in turn makes corals more susceptible to diseases, which also seem to be increasingly common as the oceans warm. The first two species to be listed as threatened under the U.S. Endangered Species Act due to climate change were elkhorn and staghorn coral, for which warmer water, diseases, and storms were given as the top three reasons for listing. Other stressors such as pollution and overfishing can sharply reduce the resilience of coral reefs to global changes. An increase in global mean temperature over preindustrial levels of just 1.5°C may be sufficient to make the persistence of coral reefs unlikely. The global loss of reef-building organisms is not unprecedented in the geological record.

## Final Thoughts

Scientific understanding of climate change and its effects on physical, chemical, and biological systems is rapidly evolving, and will continue to do so. We need to add this new information to our plans as we get it, but we cannot wait until we get it all because we will never get it all. Change will continue and we must continue to adapt. Identifying and filling key data gaps is important, but it does not in and of itself reduce the vulnerability of conservation to climate change. Documenting decline is not a solution.

# Chapter 3

# *Reconceiving Conservation and Resource Management*

There is no box.
—*Amory Lovins*

Climate change is not speculation: it is our present and our future. Integrating this reality into our thinking at all levels is a key element of effective, robust medium- and long-term planning. The climate commitment (Wetherald et al. 2001; box 3.1) is not just about atmospheric greenhouse gas concentrations, but also about the effects of those concentrations as they are manifest in changes in temperatures, precipitation, sea level, and planetary chemistry. The inertia of the climate system makes it hard to quickly change the trajectory we have created thanks to more than a century of significant greenhouse gas emissions. Bringing emissions into check in the next few decades will still result in a path that will take centuries to stabilize atmospheric concentrations of greenhouse gases, multiple centuries to stabilize temperature, and millennia to stabilize sea-level rise (IPCC 2001; see also fig. 3.1).

Adaptation and mitigation (limiting the negative effects of climate change and limiting change itself) are the yin and yang of an interconnected approach to dealing with climate change. Both are now essential for successful conservation and resource management: it is no longer either/or. Limiting the rate and extent of climate change is essential if we hope to enact effective, affordable adaptation, and we need effective adaptation because we have waited too long for mitigation alone to solve the problem. Some suggest that we can be less aggressive with mitigation and make up for it with adaptation, but adaptation can take us only so far. As a result, we need to implement

**BOX 3.1 CLIMATE CHANGE COMMITMENT**

**If atmospheric greenhouse gas concentrations stabilize today, the momentum of the climate system is such that it would take a century or two for the global temperature to stabilize and a millennium or more for sea level to do so (IPCC 2007a; fig. 3.1). This is our climate commitment, the amount of change made unavoidable by our emissions to date. Our current climate commitment is on the order of half a degree of warming and a 10-centimeter rise in sea level due to thermal expansion alone by the end of the twenty-first century (Meehl et al. 2005). Actual warming will be greater because we will not be able to stabilize greenhouse gas emissions immediately, and actual sea-level rise will be much greater due to the influence of melting glaciers and ice caps.**

adaptation, thinking of mitigation as a key step in the adaptation process; and implement mitigation, thinking of adaptation as a key step in the mitigation process. Adaptation is a way to buy time for systems (human or ecological) while we take action to limit the extent of climate change, as well as a new way of doing things within the context of our permanently altered climate.

To remain effective, we will have to shift both the culture and philosophy of conservation and resource management. Culturally, these endeavors need to become more holistic, engaging in cross-sector collaboration (between natural resource and built environment planning, for instance) and considering actions such as reducing greenhouse

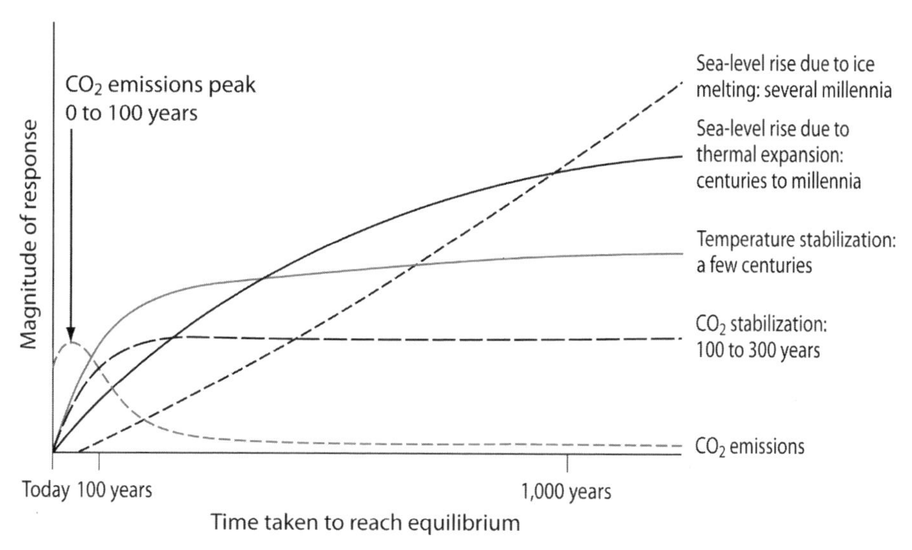


FIGURE 3.1 Time to stabilization for atmospheric $CO_2$, temperature, and sea level following stabilization and reduction of $CO_2$ emissions. After fig. 5.2 in IPCC 2001.

gas emissions as part of conservation or management solutions. Long seen as beyond the purview of natural resource practitioners, such actions are necessary for successful long-term conservation and management. Indeed, reducing greenhouse gas emissions has already been included in a handful of recovery plans for endangered species (Povilitis and Suckling 2010). This harkens back to the idea that prevention is cheaper and more effective than a cure.

Another necessary shift relates to the current prevalence of thinking about protection and planning based on historic or current patterns. With climate change, especially with the possibility of tipping points (see chapter 2), it may not be possible to maintain current conditions or return to past states. Planning goals need to change from ideals of static preservation and restoration to past conditions to something more dynamic.

The only way to avoid dangerous climate change is to limit our alterations of the climate system. While adaptation may be able to compensate for lower levels of change, its ability to compensate declines as the magnitude of change increases (see fig. 3.2). It is almost certainly cheaper to protect systems from the effects of climate change by ensuring that climate change does not occur in the first place. Unfortunately, we are now at a point where we must take on both mitigation and adaptation on a grand scale.

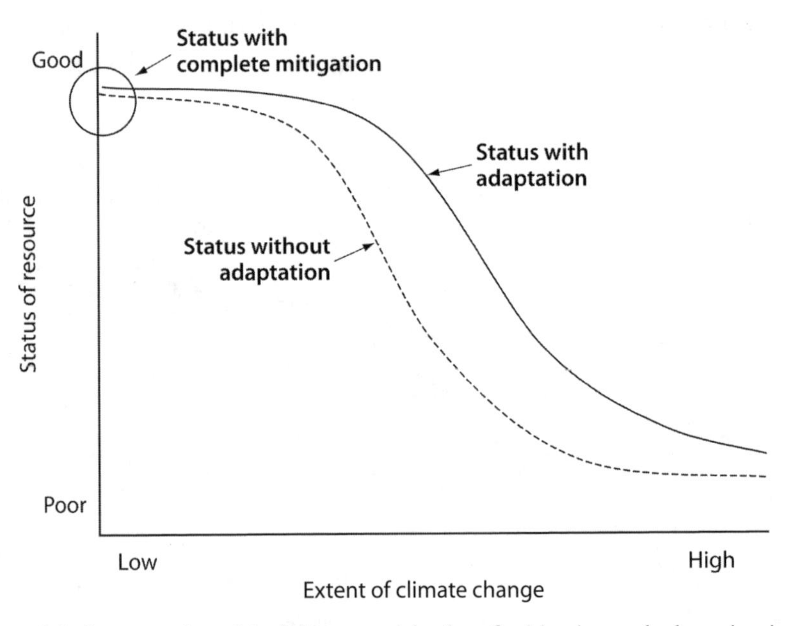


FIGURE 3.2 Conceptual model of the potential roles of mitigation and adaptation in meeting conservation or management targets. With low levels of climate change, adaptation can maintain the status of the resource of interest. As the extent of climate change increases, adaptation becomes less and less able to compensate and the status of the resource inexorably declines.

## Vulnerability of Current Conservation and Management Paradigms

Current resource conservation and management paradigms evolved during a time when the climate system was seen as having some year-to-year variability but being relatively stable over human time frames. As a result, current practices focus on space more than time, and attempt to manage for a status quo (or ideal past) world. Plans are built on the assumption that resources and habitats will continue to be available where they are now, and that the past is a reasonable guide for the future. These assumptions may not hold in the face of climate change.

### *Static Data*

The allocation of the Colorado River's water illustrates the weakness of using a static snapshot approach to make decisions about a variable system. The Colorado River supplies water to Arizona, California, Colorado, Nevada, New Mexico, Utah, and Wyoming, as well as parts of Mexico. The 1922 Colorado River Compact, and several subsequent agreements, allocates the Colorado's water among these parties. When the compact was created, negotiators elected to use the previous seventeen years' flow data as the basis for allocations. As it turns out, the average annual flow during that period was one of the highest on record and had less variability than any other period on record (see fig. 3.3). The result is that the river has been overallocated almost every year since the compact was created. In fact, tree-ring records suggest that in the past twelve centuries the Colorado River has had few periods with flows that high or consistent (Meko et al. 2007).

Some of the fastest-growing regions in the United States rely on the Colorado River for water and hydropower. Yet given that close to 90 percent of the Colorado River's flow comes from snowmelt (Christensen et al. 2004), the river and those that depend on it are highly vulnerable to climate change, and in fact the western United

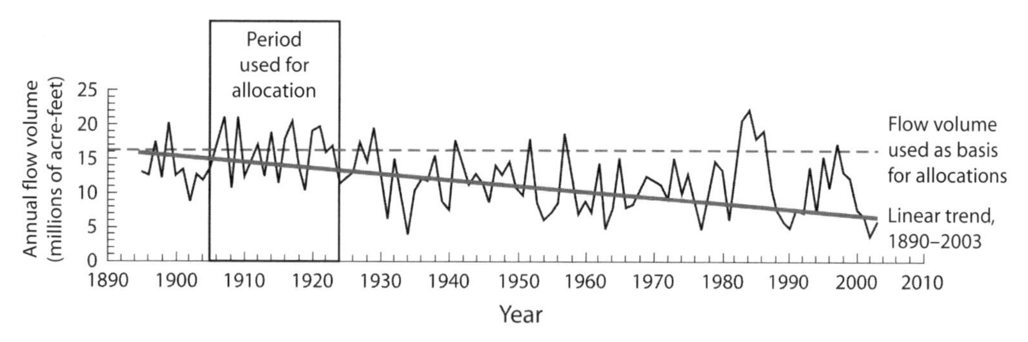


FIGURE 3.3 Estimated annual average Colorado River flow at Lee's Ferry, Arizona. The average flow rate from 1905 to 1922 was used to determine water availability for allocation. Source: U.S. Bureau of Reclamation.

States is already experiencing dramatic hydrological changes due in part to climate change. If the current climate change trajectory, water allocation plans, and use rates continue, there is a 10 percent chance that Lake Mead and Lake Powell will be dry by 2013, and a 50 percent chance that it will happen by 2021 (Barnett and Pierce 2008).

### *Static Site-Based Protection*

Protected areas are the gold standard of conservation practice. Most conservation organizations and governmental bodies base their missions and practices on protecting status quo species or habitat composition within protected areas. For example, the 1972 United Nations Educational, Scientific, and Cultural Organization (UNESCO) World Heritage Convention assigns "the duty of ensuring the identification, protection, conservation, presentation, and transmission to future generations" of key cultural and natural heritage, and calls for "an effective system of collective protection of the cultural and natural heritage of outstanding universal value, organized on a permanent basis and in accordance with modern scientific methods." World Heritage Sites, be they cultural or biological, are at risk from climate change, and protecting them in their current state over several generations may not be possible. This is particularly true for those elements of natural or cultural heritage that are defined by their location on the landscape: a static protected area will not be enough to preserve buildings or habitats when the conditions they need change beyond a certain point (see box 3.2). This brings us back to the idea that preventing harm is better than trying to fix damage that has been done. The vulnerability of these cultural and biological jewels can serve as a strong rationale to reduce greenhouse gas emissions and slow the rate and extent of climate change.

Many species protection laws, such as India's Wildlife Protection Act of 1972 or the United States' Endangered Species Act of 1973, have similarly been interpreted to support protecting or restoring habitats or species to historical conditions. In most cases, designation of critical habitat has focused on locations that currently support or formerly supported the species of concern with little consideration for where species might need to be as a result of climate change. There are fortunately indications that this approach is beginning to shift. In the United States, for instance, the inclusion of climate change in species' recovery plans has skyrocketed in recent years (Povilitis and Suckling 2010). Also, existing efforts to build networks of protected areas, such as the Protected Area Network of Parks in Europe, or to coordinate conservation and management across large areas, such as the Western Hemisphere Migratory Species Initiative, provide an excellent framework for supporting connectivity and climate refugia as species move in response to climate change.

It is not impossible to create an approach that includes spatial protection but in a manner that explicitly supports flexibility. For instance, the Convention on the Conservation of Antarctic Marine Living Resources employs an ecosystem-based management approach to protecting Antarctic marine life. The partners in CCAMLR have not made the Southern Ocean off limits; rather, they attempt to minimize risk from

### BOX 3.2 WHEN IS A KEY DEER NOT A KEY DEER?

The Key deer is federally listed in the United States as endangered. Fewer than 800 individuals remain, up from an historic low of fewer than 50 in 1957 (U.S. Fish and Wildlife Service 2008). While the National Key Deer Refuge consists of more than 84,000 acres across 25 islands in the Lower Florida Keys, three-quarters of the remaining Key deer live on just two of those islands, Big Pine and No Name Keys. The maximum elevation of these islands is just 3 meters above sea level, and deer tend to avoid habitats less than 1 meter above sea level, opting for the pineland, hammock, and human landscaping found on higher ground (Lopez et al. 2004). Big Pine and No Name Keys have less than 1,200 hectares of upland habitat between them (Ross et al. 1994).

Much of the Key deer conservation effort focuses on acquiring more land, ignoring the extreme vulnerability of this area to climate change. The combination of sea-level rise, increasing hurricane intensity, and existing human population stress makes it unclear how protecting more Key deer habitat in the Florida Keys will be a successful conservation formula over the long term. In 2005, Hurricane Wilma caused pineland vegetation mortality of between 30 and 80 percent (Ross et al. 2009a). Sea level could well be around 1 meter above the preindustrial level by the end of the century. This implies a few things for Key deer conservation. One is that habitat will be lost continually as sea-level rise continues. Indeed, sea-level rise may be more rapid than currently projected if warming accelerates the melting of glaciers and ice caps. Habitat loss is not just about direct land loss: some vegetation in preferred Key deer habitat is not salt tolerant and will be degraded both by storm damage and by saltwater inundation of soils and groundwater. It is hard to imagine what Key deer will do as their habitat disappears. There are no obvious avenues for migration, other than out of the Keys up Highway One. Assisted migration up the Florida peninsula to a point where they are out of harm's way poses the question of whether an animal is still a Key deer if it no longer lives in or near the Keys.

The draft comprehensive conservation plan for the Key deer refuge (U.S. Fish and Wildlife Service 2008) includes sea-level rise and saltwater inundation from storm surge in its list of threats. It calls for better understanding of current ecosystem function, assessing future impacts from climate change, and developing adaptive management strategies in anticipation of change.

The Key deer is one of twenty-one threatened and endangered terrestrial species in the Florida Keys. Analysis for one of these, the Lower Keys marsh rabbit, indicated substantial loss of individuals from sea-level rise of less than 1 meter and recommended dramatic new conservation efforts to avoid species extinction (LaFever et al. 2007). The loss of upland habitat from sea-level rise, storms, and development is just the type of compound challenge that climate change brings. Protected areas, which could once be designed to support minimum viable populations and act as biodiversity safe havens, are now being undermined by the external pressures of climate change.

"unsustainable practices in conditions of uncertainty" by employing a precautionary approach. Additionally, in recent years the convention has begun to explicitly address climate change in its attempts to plan for and manage krill populations and other marine life in the region (CCAMLR 2009). Another example of an organizational mission that supports flexibility comes from the U.S. National Marine Fisheries Service, the mission of which is to provide "stewardship of living marine resources through science-based conservation and management and the promotion of healthy ecosystems." This is a useful climate change adaptation model as the goal is healthy ecosystems, which one can hope will support not just the historic assemblages of species and habitat, but also the new assemblages that appear as climate change continues.

### *Status Quo Thinking*

Like the UNESCO goal of permanent protection, the general goal of conservation has been to maintain things as they are or even, in the case of early ecological restoration programs, making them as they were. The challenge of climate change is that it is changing the world around us, taking us on a seeming one-way journey toward new habitats, communities, and maybe even species. None of this will stabilize on a human timescale, and recovery to the past becomes impossible if tipping points are reached. The vulnerability of a status quo approach is easy to understand when it comes to issues like low-lying islands and sea-level rise, where even creating a whole-island protected area may not be enough for endemic species. It is equally real, though more complicated, when it comes to the fate of ecosystems such as the Amazon rainforest, where global climate change and deforestation likewise threaten the existence of an entire system.

The damage to the Amazon rainforest due to habitat destruction and fragmentation rallied conservation practitioners and citizens around the planet in a way that few other location-specific conservation issues have. It has resulted in everything from consumer movements to international governmental agreements to slow the rate of deforestation. One of the crowning achievements in this effort has been the creation of the Amazon Region Protected Areas (ARPA)—a joint effort by the Brazilian government, the German Development Bank, the Global Environmental Facility, the Brazilian Biodiversity Fund, and many other organizations—with the aim of protecting an additional one-third of the Brazilian Amazon over the course of a decade. The first phase of the project is budgeted at US$81.5 million. Unfortunately, climate change is altering the very forest parcels that are being protected. Several models suggest that by the end of the century, decreased precipitation and altered seasonal cycles could cause as much as 70 percent of the Amazonian rainforest to be replaced by caatinga or savanna vegetation (Cook and Vizy 2008; see also fig. 3.4). The irony is that in addition to its value as a home for people and spectacularly diverse plants and animals, the Amazonian forest also stores vast quantities of carbon. The forest loss caused in part by climate change will result in even more climate change as the Amazon's ability to take up and store carbon drops dramatically. By the end of the century the Amazon may hold as little as a quarter of the carbon it holds today (Cox et al. 2004).

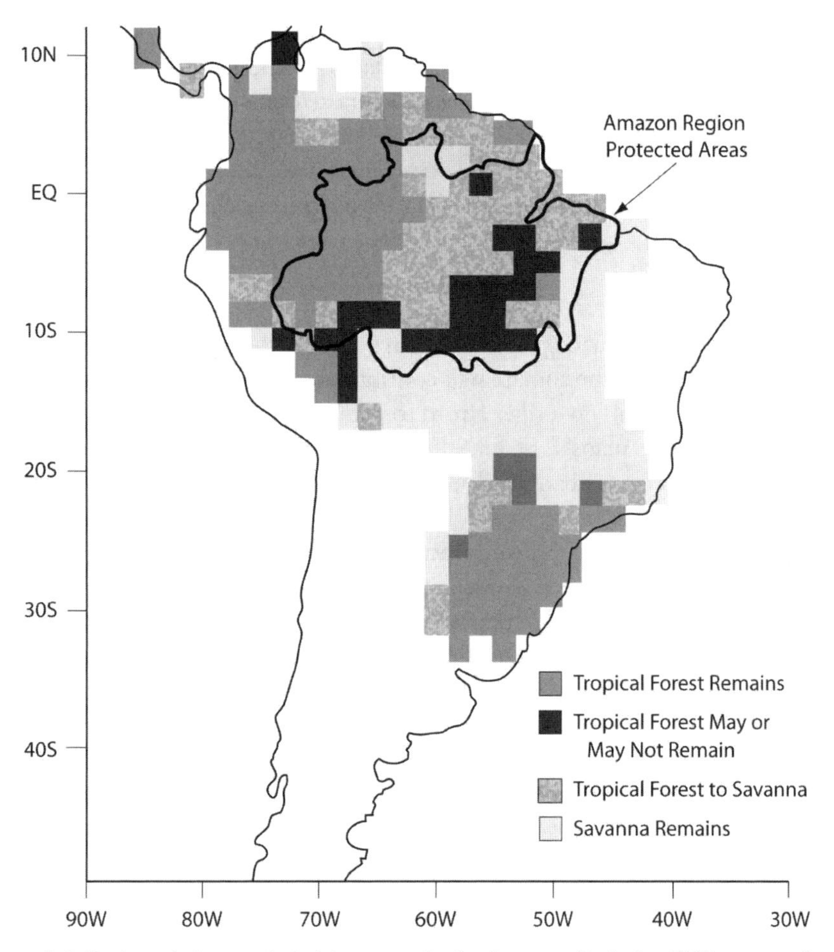


FIGURE 3.4 Projected changes in habitat types in the Amazon Basin by 2100 as a result of climate change. After Salazar et al. 2007.

Clearly there is a desire to protect the Amazon for a host of reasons, including a desire to right environmental and cultural wrongs or to maintain the function and beauty we so appreciate. However, with the potentially sweeping changes brought by climate change, we need to rethink what we want to protect and how we are going to make it happen.

## New Frame for Conservation and Resource Management

Climate change is happening now, and happening faster than initial projections led us to believe. The effects are far-reaching with no ecosystems left untouched, and will carry on long beyond our current planning horizons. What is conservation when our goal can no longer be to protect or restore a place or a species' populations to the way they were? How can we manage populations whose dynamics are shifting under our very eyes? We can no longer think in terms of safeguarding or maintaining. If the old model is spatial, static, and stuck in the status quo, it would seem that we need a temporal, kinetic, and forward-thinking model to deal with the challenges we face today.

### *Temporal*

What is conservation or management that prepares for the next ten, fifty, or one hundred years? The fact that the climate will continue to change for all ranges of planning futures means that we can no longer afford to plan for just the short term. We must develop and implement plans that take into account the reality of ongoing climate change. This applies to an urban development plan on a coastal floodplain that must account for sea-level rise as much as to a forest management plan that needs to prepare for changing precipitation, species composition, and human resource use. This means that plans cannot just consider where sea level is today. We need to consider where it *will* be and plan for it to get there perhaps sooner than models suggest. Perhaps we think about relocating populations or species, hopefully learning from past experience with this approach when it was suggested as a means of coping with other threats. Perhaps we consider triage, accepting the damage we have done as irreversible for some populations, species, or ecosystems. Even now we do not work equally hard for all possible conservation targets, carrying out an ad hoc form of triage but without a broader guiding framework.

### *Kinetic*

Building the future into today's planning and management means more than using a single projected future climate in our plans. We need to plan for ranges of possible futures and ranges of acceptable outcomes, because climate change brings increased climate variability and decreased certainty about what our future will look like. Just as the data used for the Colorado River Compact came from a period of particularly high flow, it was also a period of particularly stable flow. We are no longer managing for best-case scenarios: we need to employ the precautionary principle to prepare for worst-case scenarios under both feast and famine conditions. That means preparing for both flood and drought, or whatever range of extremes a system is vulnerable to, because the likelihood is that both extremes are now more possible. It is possible that nearer-term extremes may be the Achilles' heels that critically damage systems before the long-term trends come to pass. This may be a cliché, but change *is* the only certainty.

We will need a new evaluation of how we think about nonnative species. Species, both endangered and cosmopolitan, will be shifting into new areas that may themselves be managed to protect other species, resources, or ecosystem services. This may create conflicting management requirements, or it may require new philosophies about how and what we manage. Are some nonnative species now climate refugees that need protection and cultivation rather than eradication? While we deal with what may become labeled as an explosion in invasive species, we will also have to prepare for the likelihood that the number of threatened and endangered species will be increasing.

### *Forward-Thinking*

Given that existing species assemblages may not remain intact as species respond at different rates and in different ways to climate change, we may need to start thinking about how to protect land- or seascapes so that they are functional regardless of who lives there. We may also need to clarify what we mean by functional (see box 3.3).

The new frame for conservation and resource management is as much about philosophy as it is about methodology. It will be crucial to imbue all of our decisions (research, management, communications, and policy) with clear recognition as to the role

#### BOX 3.3 MAINTAINING FUNCTION IN THE FACE OF CLIMATE CHANGE

Climate change poses many seemingly insurmountable obstacles: surpassed temperature thresholds, loss of sea ice, disconnected phenology, and land loss to sea-level rise. However, these have all happened during past periods of climatic change, albeit generally at a slower rate, and we can look to the past to see how systems have or have not adapted. A common response to sea-level rise, provided it was not occurring too quickly, has been for coastal species and ecosystems to move inland with the rising seas. This is not possible if the land behind the current coastline is degraded or obstructed. In North Carolina's Ablemarle Peninsula, the Nature Conservancy, U.S. Fish and Wildlife Service, and local partners have a project under way to preserve three elements of this coastal ecosystem as sea-level rise progresses: hydrology, wetlands, and oyster beds. For two centuries these lands have been the site of forestry, agriculture, and pest eradication. Extensive drainage ditches were cut to support these endeavors in this wetland peat environment . As sea level rises, these channels expedite saltwater inundation of previously freshwater habitats. Therefore, one of the first priorities is to create a system of water flow less conducive to saltwater inundation. A secondary benefit of limiting saltwater intrusion is that the carbon in the peat soils is not released. The next priority in preparing the inland habitat to become the new coastal salt marsh is planting native flood- and salt-tolerant species. Finally, to protect this coastal habitat from storm surge and erosion, native oyster reefs will be established in the near-shore shallow habitat. The goal of this project is not to protect the system as it is, but to create a functional coastal salt-marsh system that is resistant and resilient to climate change.

climate change will play in their success or failure, and to incorporate uncertainty about the future climate into our planning. Robust and effective conservation and management that make the most of our limited workforce, time, and funds will require that we start to consider how today's management actions can lead to effective conservation in a very different tomorrow. We can view these challenges in a way that inspires creative new thinking and catalyzes action, avoiding paralysis or denial without downplaying the magnitude of the challenge. How do we begin to incorporate this new framework into our thinking? These issues are further explored in chapter 4.

### *How Current Climate Change Is Different from Past Climate Changes*

Natural systems certainly adjust and adapt to changing conditions, and the world did not in fact end during previous periods of rapid or dramatic climate change. However, such periods *were* associated with enormous changes in plant and animal communities, including some episodes of mass extinction. Change is currently happening faster than during many past climate shifts, which in and of itself would suggest significant ecological effects. Today's climate shift is also happening after a long period of very stable climate, and it is happening across land- and seascapes that have been dramatically altered by human activity. These alterations, due to a growing human population with a growing demand for resources, both contribute to climate change and create concomitant challenges that further diminish the ability of natural systems to adapt to that change. These additional challenges are the traditional list of environmental stressors that led us to the biodiversity crisis: habitat destruction and fragmentation, overharvest, pollution, and invasive species, among others.

## Final Thoughts

The permanence of climate change is often hard to grasp from a human planning perspective. Our success in conservation and sustainable resource management, however, requires that we come to terms with this new and changing set of circumstances and understand that they will continue to change for the foreseeable future. We need to develop and employ a more flexible, responsive approach than those we have historically embraced.

# PART I

## *Building the Plan*

The problems climate change poses for resource management and conservation should be clear by now. The prudent course is clearly to take action to limit vulnerability, but this is not always what we do when confronted with a challenge. The next four chapters focus on how to break out of paralysis and move toward building and implementing a plan for incorporating the reality of climate change into what we do, increasing the likelihood that our work will be successful. In essence this is creating a sort of climate change insurance plan for our work.

Successfully adapting to climate change is part philosophy, part science, part governance and management, and part having clear goals. This combination can help overcome the barriers of uncertainty and indecision that seem to crop up whenever climate change planning is on the agenda.

Starting with a philosophical framework (chapter 4), this section offers assurance that successful adaptation does not mean getting it right the first time or achieving absolute knowledge and certainty. Adaptation is more about looking at the world differently and learning as we go. This way forward is not completely random, however: being climate savvy has a few basic tenets that can help guide early efforts. These tenets include rejuvenation of traditional approaches (protecting adequate and appropriate space and reducing nonclimate stressors), modification of familiar themes (managing for uncertainty), and new considerations that have not generally been seen as part of conservation and resource management (reducing local climate change and addressing global climate change).

After becoming philosophically grounded, the next step is to understand the vulnerability of our own work or of particular projects to climate change (chapter 5). We can do this in a variety of ways depending on budget, expertise, and other factors, provided we follow a few best practices. Chapter 6 explores how to move from assessing vulnerability to developing adaptation plans. This does not necessarily mean discarding existing plans and processes: it may simply mean explicitly building climate change and its effects into those plans and processes. Developing adaptation plans includes some concepts with which we are all familiar (accounting for human behavior and values, stakeholder engagement, and monitoring) as well as a few newer ones (resistance and resilience, and robust decision-making). While some people are intimidated by the concept of climate change adaptation, it is not anything magical or beyond the ken of ordinary mortals. It is simply a means of maximizing the likelihood of continued efficacy for your work in the face of climate change. It increases the value of your investment.

Not to be overlooked is the need to be clear about our underlying vision, goals, and objectives. These may need to be adjusted to become more robust to climate change, but we should be cautious about changing them wholesale in response to a perceived climate change gold rush.

Models provide a framework around which to build adaptation plans (chapter 7). They may be conceptual or computational, simple or complex, and are a means of synthesizing existing knowledge about climate systems or biological or social responses to climate change. Models can be an excellent vehicle for exploring plausible scenarios for the future and testing management strategies and policies under these changing conditions. It is important to understand their benefits and limitations so that their output can be used appropriately.

Above all, avoid looking for excuses not to act. Enabling conditions are never perfect. When King County, Washington, decided that addressing climate change in their county plan was important for the protection of county investments, they simply started doing it. Even when essential actions such as a levee system modification fell outside their direct mandate, county officials worked to make improvement happen rather than bemoaning their limitations.

When stranded and starving on a tropical island with no machete to cut open coconuts, the choices are starvation or finding another way to get the coconuts open. Get out there and use what you have to create a plan that will work while you continue to learn.

# Chapter 4

## *Buying Time*

### THE TAO OF ADAPTATION

> Well . . . the world needs crazy ideas to change things, because the conventional way of thinking is not working anymore.
>
> —*Alexis Ringwald*

There is a point between accepting the need to adapt their work to climate change and actually doing it where many people falter. Making this leap means figuring out exactly how to adjust what you do to this new way of thinking, and like all changes it can seem onerous. Here we present an adaptation perspective that we hope will help you to find a path forward.

Adapting to climate change is not just a scientific problem or a new methodology: it is a philosophy that must imbue all of our activities and decisions. In some cases it presents an opportunity to finally start acting in ways we have wanted to but never did—a fuller realization of idealized conservation philosophies. It may require changing goals or approaches, or just adjusting our outlook. In all cases, it provides an opportunity to realize some of the very basic aspirations at the heart of our work by giving them a new context.

A key aspect of adaptation is hedging our bets as we manage for an uncertain future. Even with the best science and the best predictive tools, projections of future climate change and its effects will never be definitive. Setting aside the uncertainty inherent in models, be they climate, ecosystem, or what have you, so much depends on the choices we humans make (e.g., whether and how much we reduce emissions or

### BOX 4.1 MALADAPTATION

Not all responses to climate change will reduce our vulnerability. Some suggested responses to climate change are not themselves robust to the effects of climate change, and others increase the vulnerability of species or ecosystems to change. For instance, hydroelectric, nuclear, and biomass production are all touted as climate change mitigation options because of their lower carbon intensity. Yet all are vulnerable to climate change. During the 2003 summer heat wave in Europe, many of France's nuclear power plants had to be temporarily shut down because the combination of high air and water temperatures and limited access to water meant the plants could no longer be operated at safe temperatures. While hydropower can provide large amounts of energy with relatively low greenhouse gas emissions, big dams make many species that depend on the river much less resilient to change.

Other attempts at planning for adaptation can actually make climate change worse. Some guidance documents suggest bottled water and air-conditioning as ways to cope with extreme heat. From a disaster management standpoint, these might seem like fine suggestions, but from the perspective of sustainability and alleviating global climate change, they are dreadful. Both air-conditioning and bottled water have intense carbon footprints, only making matters worse in the long run.

how we respond to changing conditions), a factor that can never be predicted definitively. As a result, the best approach is to establish plans that account for this uncertain future and respond flexibly to changing conditions. Consider which actions make your ability to meet your underlying goal most robust given an array of future scenarios, including scenarios of human and ecosystem responses as well as of climate change itself. Sometimes the actions you can take to reduce vulnerability are quite similar across a range of possible futures.

### BOX 4.2 TENETS OF ADAPTATION

- Protect adequate and appropriate space for a changing world
- Reduce stressors that are exacerbated by or exacerbate the effects of climate change
- Manage for uncertainty
- Reduce the rate and extent of local and regional climate change
- Reduce the rate and extent of global climate change

No single element or component of adaptation is a solution on its own, and there is no universally best set of tenets. Successfully adapting to climate change relies on a mixture of approaches as well as perpetual review and modification as new information comes to light, new ideas are generated, and additional changes take place. What we present here is a set of tenets that has worked for us as a conceptual framework for developing adaptation plans. We describe them briefly as a point of reference for the remainder of the book, but encourage readers to develop their own framework for adaptation if these do not resonate for them.

## Protect Adequate and Appropriate Space for a Changing World

As discussed in chapter 3, protected areas are vulnerable to climate change: they remain fixed in space while environmental conditions, species, and even whole biomes move. Despite this inherent vulnerability, protection of physical space can still play an important role in adapting conservation and resource management to climate change provided it is done in a climate-savvy way.

One approach is to prioritize protection of places that are likely to maintain more stable climatic conditions, often referred to as climate refugia. Potential refugia may be identified using historic data to look for locations or types of locations that have changed more slowly during the last century or during previous periods of climate change. In marine settings, for instance, areas with strong currents and mixing due to oceanographic or geologic features may serve as temperature refugia. However, the transient nature of some of these sites, especially during El Niño years, and new findings that upwelling waters often have reduced pH (Feely et al. 2008) make this less of a sure bet. On land, refugia may be found in regions with high topographic variability. It is important to recognize that even locations that have been refugia in the past have some upper threshold at which they cease to be refugia.

If protection of particular species is the goal, supporting the possibility of range shifts may require very large multi-use protection regimes, networks of protected areas, corridors, or movable protected areas that track species over time. Incorporating climatic gradients and refugia, habitat heterogeneity, and connectivity across landscapes may help. Such concepts are not unique to climate adaptation, and are already included in conservation schemes such as the Yellowstone to Yukon Conservation Corridor and the many ridge-to-reef or whitewater-to-bluewater protection plans (fig. 4.1). Some protection could also focus on populations of the focal species likely to be better adapted to future climates, such as populations currently living in warmer areas.

Protected areas may also be designed around maintaining ecosystem functionality. This could mean the continued existence of a wide diversity of plants and animals, or the continued existence of the benefits of nature on which people rely, such as water filtration or flood control. Climate change necessitates thinking forward to what functions will be possible to maintain in particular locations, and what will be needed to maintain them in light of changes in systems as basic as water or carbon cycles.

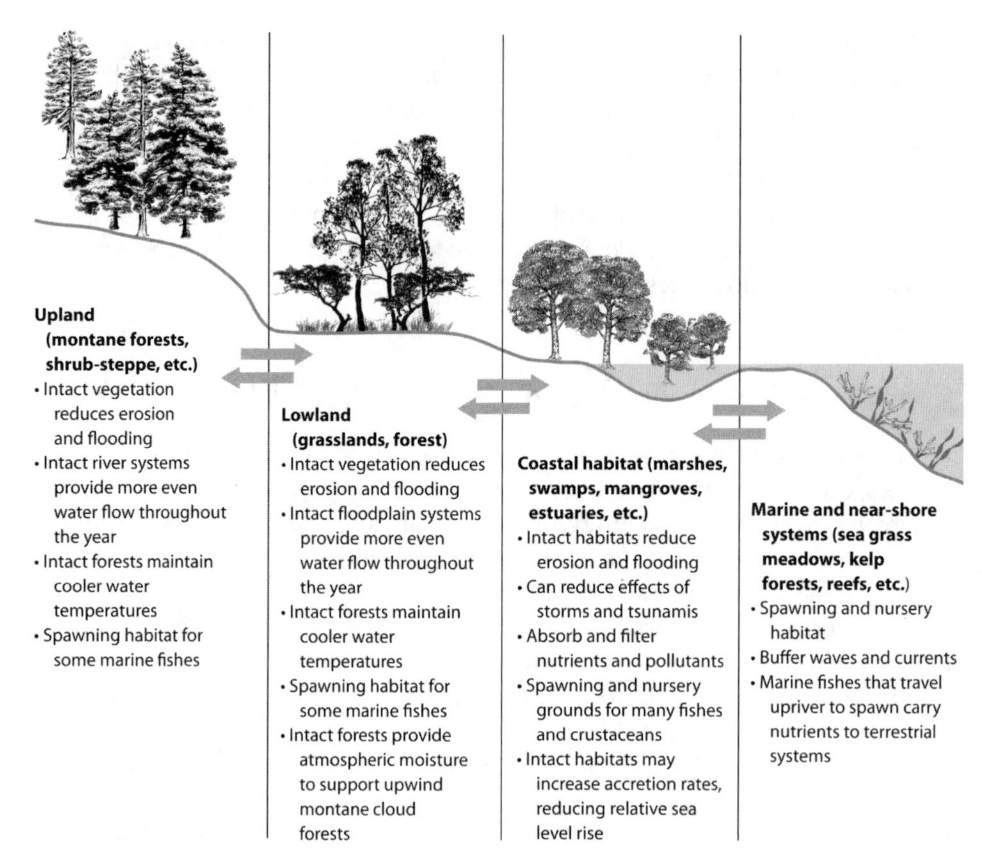


FIGURE 4.1 Schematic of ecological connections between land and sea. Upland changes in land cover, pollutant input, or other factors can have dramatic effects on the health of coastal and marine habitats, and species such as salmon that use both marine and terrestrial habitats are sensitive to changes in either.

Restoring or maintaining function by focusing on historic composition may no longer be an option.

We need to think of spatial protection not in terms of protected "areas" but in terms of an integrated functional matrix of areas spanning a range of uses and states of naturalness. We cannot protect some places from climate change and ignore others. Rather, we need a model whereby we manage and protect the whole, not the parts. Protected areas, species protection, and connectivity are explored in more detail in chapters 8, 9, and 10.

## Reduce Stressors That Interact Negatively with Climate Change

Some of our greatest conservation successes have had little to do with the creation of protected areas and everything to do with limiting environmental stressors such as pollution or overharvest. Bald eagles were pushed nearly to extinction by dichlorodiphenyltrichloroethane (DDT), but have recovered spectacularly following a ban on DDT in the United States and Canada (Grier 1982).

There are many potential interactions between the stresses of climate change and these other stresses. Climate change may exacerbate the adverse effects of other stressors, or other stressors may worsen some negative effects of climate change. In a handful of cases, climate change and other stressors may ameliorate each other's negative impacts. Almost all existing environmental stressors have the potential to interact with climate change in some way.

### *Pollutant Stress*

The ubiquitous nature of pollutant stress has been recognized for decades, but climate change is poised to compound the problem. Changes in temperature or chemistry of soil and water can lower the tolerance of plants and animals for pollutants, or increase the toxicity of pollutants. These changes, as well as changes in flood and drought cycles, can also increase exposure to or availability of pollutants in an ecosystem. For example, polychlorinated biphenyls (PCBs)—a widespread class of toxic chemicals linked to wildlife die-offs, human health problems, and developmental issues in babies whose mothers were exposed to high levels of PCBs during pregnancy—become even more toxic with increasing temperature for some species (Fisher and Wurster 1973). Although PCB production has been banned in much of the world for decades, PCBs stored in glacial ice are being released as glaciers melt, leading to increasing levels of this contaminant in some areas. The effects of pollutant/climate interactions may be manageable if we accept and act on the reality that threshold or regulatory levels may need to be adjusted as climate change alters environmental conditions. These issues are more thoroughly explored in chapter 14.

### *Pests, Diseases, and Invasive Species*

Climate change can likewise compound the problem of nuisance species in several ways. Higher temperatures allow some pest species to expand into new regions or dramatically increase their population growth rates, and plants and animals stressed by changing climatic conditions may become less able to compete with pest species. Some dramatic cases involve bark beetles in northern forests where millions of hectares of pine forest have been killed by pests that are native to the region but were historically kept in check by cold winter temperatures. In other regions new pest species are moving in, leading to calls to increase pesticide use and thus compounding the problem of environmental pollution. Less damaging approaches to pest control such as increasing natural resilience may also prove useful. In the case of some forest beetle infestations, for instance, restoring natural fire regimes and encouraging a diversity of tree ages and species may help to reduce the severity of beetle outbreaks. This illustrates the importance of considering the longer-term, follow-on effects of potential adaptation options as well as the direct, immediate ones.

Invasive species management, already an enormous challenge, will not be made any easier by climate change. Most invasive species are cosmopolitan, meaning that they can tolerate and even thrive in a wide array of conditions, giving them an advantage over those native species that are more tightly adapted to historic local conditions. Thus changing conditions brought on by climate change could give some invasive species an extra boost. Anticipating this challenge allows managers and regulators to step up efforts to keep invasive species out of areas where they do not currently occur, or to prioritize efforts to combat them where they already have a foothold. A second challenge, however, will be deciding when to consider a new arrival a climate refugee to be accepted rather than a threat to be gotten rid of. Interactions between climate change and invasive species, pests, and disease are discussed in chapter 12.

### *Overharvest*

Overharvest or overexploitation (of forests, fishes, water, and other resources) has the immediate effect of reducing the availability of the resource being exploited, and in the longer term can damage ecosystem function (e.g., loss of keystone species) or the availability of linked natural resources (e.g., decreased water availability following widespread deforestation). Climate change alters how quickly or completely resources replenish themselves, making it more difficult to predict sustainable harvest levels. For instance, a particular fish population may grow more quickly or more slowly as a result of changing temperatures, and the replenishment of groundwater supplies will be affected by changing rainfall and evaporation. Additionally, overharvest reduces the genetic diversity that helps populations cope with changing environmental conditions both in the here and now and on an evolutionary level. The issue of overharvest is discussed more thoroughly in chapter 13.

## Manage for Uncertainty

Many people attribute their lack of action on climate adaptation to a feeling that there is still too much uncertainty to make an informed decision. This is not true. We make decisions based on incomplete information and under uncertain conditions regularly, as indeed we must. When decisions have to be made (allocating water resources, for instance), an informed decision is one based on all existing information, including information about what we do not know or are uncertain about. When uncertainty is high or key pieces of information are missing, we can implement actions that allow us to adjust our course as more information is gathered, or prepare for a variety of different potential outcomes. This is true for deciding whether to wear a raincoat when leaving the house and whether to buy or sell something now or wait for the price to change. It is true for developing recovery plans for endangered species. It is also true when it comes to responding to climate change. We acted despite not knowing the future with certainty before we were aware of climate change, and there is no reason to wait for a firm map of the future now that climate change is part of our reality. Rather, we must learn to manage for uncertainty, to make decisions that include uncertainty as another piece of information, not a lack of information.

Climate change will include linear or steady changes, but it will also include abrupt changes or thresholds and greater variability. While many climate projections focus on what the climate will look like in 50, 100, or 1,000 years, the path from now to then will almost certainly be a roller coaster rather than a straight line of change. Planning for the next 100 years means planning for that roller-coaster ride as well as for conditions at the end of the ride. Two approaches to managing in the face of uncertainty include scenario planning (when uncertainty is high and controllability is low) and adaptive management (when uncertainty is high but controllability is high as well). Scenario planning allows planners and managers to explore the effectiveness of various strategies across a range of plausible futures. Adaptive management puts management actions into an experimental framework. Although the term is now often used simply to indicate a need to adjust plans at some later date, true adaptive management plans specify what information will be gathered to evaluate management success, and how and when it will be used to adjust management actions. Both are dealt with in greater detail in chapter 16.

There are of course myriad other ways to address uncertainty, including use of the precautionary principle, bet-hedging, and risk-management approaches. With any of these strategies for coping with uncertainty, managers and planners must regularly assess whether or not they are working to achieve the underlying goal. All approaches can result in poor decisions if the process rather than the outcome becomes the focus.

At a higher level, we may need to adjust governance structures to allow for the flexibility and responsive learning needed to cope with the effects of climate change. Many current regulatory and management laws and policies are built on a fairly rigid command-and-control model. For better or worse, this makes it more difficult for organizations and individuals to make changes in how they do their work. On the one hand, it makes it more difficult to erode existing environmental protection. On the other hand, it makes it more difficult to alter regulations as it becomes clear that existing protection is not sufficient, for instance reducing allowable pollutant levels as climate change increases the toxicity for particular pollutants. An alternative to command and control is to develop so-called agile institutions (Tano 2006). This model includes six key dimensions (robustness, resilience, responsiveness, flexibility, innovation, and adaptation) with the final goal of decreasing institutional and decision-making complexity. Adaptive governance is discussed further in chapter 16.

## Reduce Local and Regional Climate Change

Even while the world is changing around you, it is possible in some cases to reduce change in the local or regional climate. Options may include reducing deforestation (locally or in areas that supply moisture and water for your region), restoring riparian

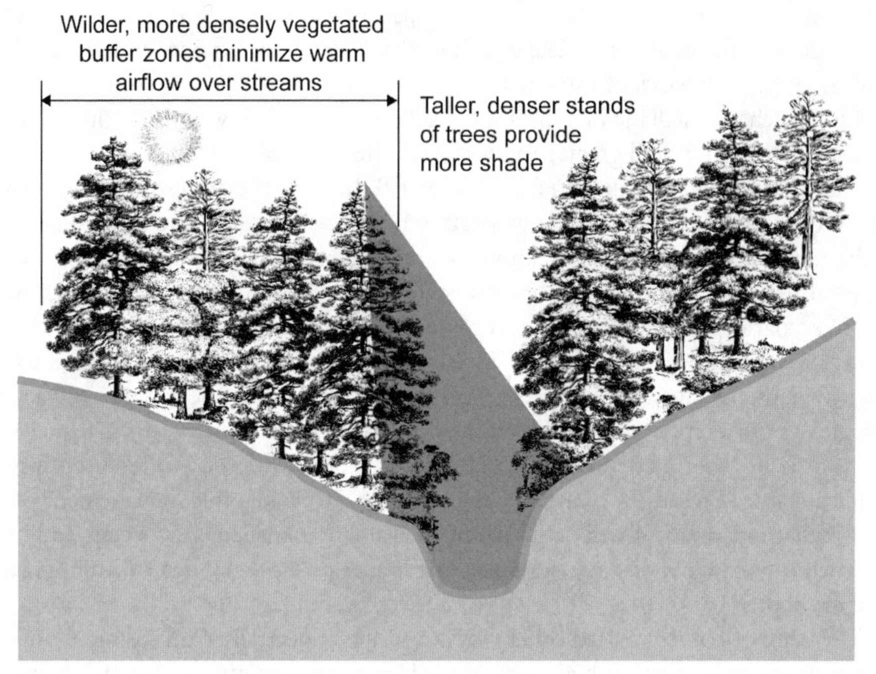


FIGURE 4.2 Influence of riparian vegetation on water temperature. Tall trees and wide areas of dense vegetation around rivers and streams can keep water cooler and reduce evaporation.

and coastal vegetation that keeps water cool (fig. 4.2), or decreasing the heat island effect of urban areas by reducing the area of dark surfaces, increasing shading, or increasing reflective surfaces.

Large forests, such as the Amazon Basin, support their own climate. Significant clearing of trees can make the area warmer and drier, or subject it to greater temperature extremes. In other cases, forests in one location are essential to maintaining the climate or water supply in another. For instance, extensive clearing in Rwanda's Gishwati forest has been linked with the disappearance of several streams and freshwater sources outside the forest. The drying of Costa Rica's Monte Verde Cloud Forest is due in part to lowland deforestation as well as to climate change. There is even a link in the hydrological—and thereby climatic—conditions in the Congo and Amazon forests, so that protection or degradation of either of these forests affects the other.

Replanting vegetation along coasts, rivers, and streams can have not only impressive effects in terms of direct shading and cooling, but more generalized cooling effects perhaps due to increased water retention in the system (see box 4.3). For freshwater systems, maintaining higher flow or water levels can help keep water temperatures lower. One idea being floated in the American west is the reintroduction of beavers. These dam-builders disappeared from much of their original range due to the fur trade or calls by some to exterminate them for their tendency to engineer areas to suit their own desires rather than those of developers and farmers. In areas where

### BOX 4.3 CARLOS DREWS, SEA TURTLE PROPHET

For six years, Dr. Carlos Drews has designed and adjusted conservation strategies in Latin America and the Caribbean to protect marine turtles from a suite of ills: overexploitation for eggs, meat, and shells; bycatch in hooks and nets; light pollution that disorients hatchlings and mothers; and habitat destruction. The population of adult hawksbill turtles in the nineteenth century was roughly 11 million, dwindling to fewer than 30,000 today (McClenachan et al. 2006). Despite the tremendous reduction in population numbers, Drews felt that progress was being made. Then he started to think about how climate change would affect turtles and their habitat, and worried that decades of conservation investment could be wasted. Most obvious was the loss of nesting beaches to sea-level rise. While beaches naturally shift inland with changes in sea level, this was no longer an option where human development (roads, hotels, cities) had sprung up directly behind beaches. Additionally, there was the fact that sea turtle gender is determined by the temperature at which eggs incubate. The warmer the sand around the buried nest, the greater the number of embryos become females. This may be fine for a few generations, but as males become fewer and farther between, the population's reproductive viability will falter. If the sand gets too warm, embryos simply die before hatching. Sea turtles also rely on ocean currents to help with their long travels between feeding, breeding, and nesting grounds. Climate change is changing winds, ocean salinity, and ocean temperature, which are then changing currents and could disorient the turtles. Finally, the feeding grounds of turtles—sea grasses and coral reefs—are being adversely affected by climate change themselves, becoming less productive and in some cases disappearing.

These changes required Dr. Drews to rethink his sea turtle conservation paradigm. He set out to understand how beaches may be flooded or reshaped by rising seas and whether their ability to move inland as the sea rose was blocked by development, cliffs, or other factors. His new paradigm also included working to maintain local climate conditions on the beach, keeping temperature increases at bay. To do this, he investigated the effects of restoring beach vegetation. He found that this vegetation decreased the temperature not only under the areas that it directly shaded, but more extensively across the entire beach. Both the sea-level rise and temperature adaptation strategies include a monitoring component so that Dr. Drews can gauge the ongoing effectiveness of his work and make additional modifications as the climate continues to change. This also allows him to better share the lessons and mechanisms of his approaches with other sea turtle conservation efforts. The larger issues of currents, reefs, and sea grass beds are still in development.

beaver populations are bouncing back, the increased water retention and groundwater recharge facilitated by their dams is helping to maintain favorable stream conditions for aquatic and riparian species despite changes in precipitation patterns.

More extreme examples of limiting local climate change include actions in the European Alps and the Great Barrier Reef Marine Park. In Switzerland, an effort is being made to slow the melting of a glacier popular for skiing—a strong economic and cultural driver in the region—by wrapping it in highly reflective white plastic during the summer. On the Great Barrier Reef in Australia, coral reefs are bleaching because of increasing water temperatures, perhaps exacerbated by periods of particularly high penetration of sunlight into the water or other stressors. Individual dive concessions are experimenting with directly shading reefs with floating cloths and using sprinklers to disrupt the water surface and decrease the intensity of solar ultraviolet radiation reaching corals. Such efforts are expensive and time-consuming, and in both the Swiss and Australian cases there are extensive tourism industries as well as cultural identities supporting and driving these actions. It is not clear that there would be funds for such extreme approaches under many conditions. There has, however, been reasonable funding for less drastic local climate control approaches like planting trees in cities to increase shade and reduce urban heat island effects.

## Reduce the Rate and Extent of Global Climate Change

There is a limit to what can be accomplished by adaptation and efforts to reduce local or regional change. All systems, or their component parts, will eventually reach a breaking point where there is too much change for their continued existence in a given location, or change happens too fast for them to move, adapt, or adjust. One simple assessment of existing information on the temperature tolerance of a variety of species indicated that even a 2°C increase in temperature relative to preindustrial times would be hard to accommodate through adaptation, and greater increases would soon make effective adaptation impossible (table 4.1).

As a result of the inherent limits of adaptation, a key element of successful adaptation will be successful mitigation of climate change. We must limit the amount of climate change and the rate at which it happens if we are to continue to enjoy the diverse plants, animals, and ecosystem services we currently have.

For some species, such as those dependent on Arctic sea ice, limiting the rate and extent of climate change may be the only adaptation option. In the last several years three species have been listed under the U.S. Endangered Species Act as threatened due to climate change, and several more are proposed for listing. These species live in the areas where climate change is happening most quickly (high-elevation mountaintops, polar regions) or in tropical areas where species are not accustomed to much climatic variability. The most famous of the listed species, the polar bear, is reliant on sea ice for food and reproduction. Climate change is causing the rapid loss of this habitat, with the area of permanent ice and the time that seasonal ice is around getting shorter each

TABLE 4.1 Assessment of adaptation options at various temperature thresholds over preindustrial levels (adapted from C. Parmesan survey of existing literature pers. comm.)

| *Temperature change (°C)* | *Effect* | *Adaptation options* | *Efficacy/cost* |
|---|---|---|---|
| 2 | Species lost | Some | Some successful |
| 4 | Many species lost | Few | Questionable and extremely expensive |
| 6 | Dire | Virtually none | Ineffective and exorbitantly expensive |

year. Most of the Arctic is expected to be ice-free in the summer by 2030, possibly sooner. Without ice to raise young and hunt for high-energy food such as seals, polar bears will either have to change dramatically—a process that polar bear biologists say will not happen fast enough—or disappear. Like the Key deer discussed in chapter 3, it is hard to imagine the survival of polar bears if the habitat on which they rely disappears.

### BOX 4.4 NICK LUNN: POLAR BEAR BIOLOGIST OR POLAR BEAR HISTORIAN?

Dr. Nick Lunn of Environment Canada jokes halfheartedly that he does not want to be a polar bear historian. He has been studying the ecology of polar bears in Canada for twenty-eight years, including an ongoing, long-term research program in western Hudson Bay that began in 1980. During this time, the number of bears in the western Hudson Bay population declined from around 1,200 in 1987 to 935 in 2004 (Regehr et al. 2007). The sea ice in this region breaks up three weeks earlier than it did only thirty years ago due to a warming climate, making it harder for polar bears to access the ice-dwelling ringed seals on which they rely for energy and building up fat stores. This has resulted in declining body conditions, particularly in pregnant females who must reach a critical weight to produce cubs, as well as declines in reproduction, and survival rates of cubs, juveniles, and old bears (Regehr et al. 2007). The loss of sea ice is the most critical concern with respect to polar bears. Highly specialized species in very climate sensitive habitats, such as the polar bear in the Arctic, are particularly vulnerable to environmental changes. Recent modeling suggests that in regions of seasonal ice such as Hudson Bay, the extirpation of polar bear populations by midcentury is likely (Amstrup et al. 2007). That is when Dr. Lunn becomes a polar bear historian.

The other two listed species are the Caribbean staghorn and elkhorn corals, which have seen heavy losses due to increasing water temperatures, disease, and storm intensity. While management actions are being implemented and planned to help the reefs withstand the further effects of climate change, there is a growing list of threats, including ocean acidification. As warming tropical waters push corals poleward and increasing acidity of cooler water pushes corals back toward the equator, it seems that limiting both of these stressors is necessary if we hope to have coral reefs in the future. Reefs are stunningly beautiful and diverse places that also act as nurseries for fishes and other species from outside the reefs, and provide a physical barrier that helps to protect coastlines from storm surge. In fact, after the 2004 tsunami in Southeast Asia, coastlines and communities inland of coral reefs or mangroves fared better than those not afforded this coastal protection.

The American pika, a small mammal typically found at high elevations in the American west, is emblematic of the plight faced by many high-mountain species. Because it is easily noticed through its unique call and construction of obvious, vigorously defended hay piles, historical collections provide ample location-specific specimens from the latter nineteenth and early twentieth centuries. Pika bones are particularly well preserved, adding to the strength of historic range records. Unfortunately, pikas are rather heat-intolerant. In the Great Basin of the United States, pika populations are disappearing entirely from some warmer and drier sites across the basin, and moving upslope or into cooler microhabitats in others. Because of their heat intolerance, they typically cannot move from their mountaintop islands to other mountaintop islands as this would require traveling through hot low valleys. Strangely, the pika also seems to be rather cold-intolerant. Because snowpack has declined over the last several decades throughout much of its range, some sites are no longer well insulated from winter cold—an especially important energetic challenge for a montane species that is active year-round. Thus, populations are blinking out not just because of hotter summers, but also because they are exposed to more cold in the winter. As these climatological shifts continue, pika populations appear likely to dwindle further. As with the polar bear, one of their best chances for survival in the wild is to limit climate change so that some suitable habitat remains.

### *Learning to Live with Climate Change*

For some human communities, limiting climate change is likewise the best bet for continued availability of food, water, and other of nature's services on which they and their way of life depend. Subsistence communities in the Arctic are finding hunting and fishing more difficult due to disappearing lakes, diseased salmon, and thawing permafrost and sea ice. Less predictable rains are making farming an unreliable source of food and income across much of the globe. Even the land itself is disappearing: a 1-meter rise in sea level, likely by the end of the century, would make many low-lying communities uninhabitable. In the Ganges Delta, entire islands have already disappeared under rising waters. Limiting greenhouse gas emissions would slow the rate of rise and give us centuries rather than decades to move people and infrastructure out of harm's way.

Availability of freshwater for drinking, agriculture, and industry is another major issue. Water stress may be due to more frequent or extreme droughts, such as the one that killed hundreds of thousands of cattle, other livestock, and iconic wildlife species such as elephants and hippopotami in Kenya in 2009. In addition to the immediate crisis of starvation, the loss of cattle threatens the traditional culture and lifestyle of nomadic peoples such as the Maasai. Disappearing glaciers are another cause of water stress: the retreat of glaciers in the Andes Mountains has already caused a 12 percent drop in the water supply for the arid coastline that is home to 60 percent of Peru's population, and other areas are likewise at risk. La Paz, Bolivia, gets 30 to 40 percent of its water from glacial runoff, while Quito, Ecuador, relies on river basins that receive about 50 percent of their flow from glacial or snowpack runoff. In the Eastern Himalayan Mountains, glaciers that mountain communities rely on for hydropower and to irrigate their fields are projected to disappear or become insufficient to meet current needs in the next few decades (Kehrwald et al. 2008).

In mid-latitudes, it is the loss of yearly snowpack that matters. One of the primary reasons the State of California became a leader in setting emissions reduction targets is the vulnerability of that state's water supply. Roughly 35 percent of the state's yearly usable surface water comes from melting snow from the Sierra Nevada Mountains, but warming winters mean that more precipitation is falling as rain rather than snow. This means less water is stored as snow in the mountains in winter, so less water is available as snowmelt in the summer. Summer water shortages are compounded by increasingly longer and drier dry seasons. Although less severe, this same problem is playing out even up into Washington State and British Columbia.

All of these issues—loss of water, land, and food security—have serious implications for civil strife and national security. While the Colorado River Compact has left a legacy of so-called water wars in the western United States, it is possible to imagine real water wars emerging between countries as they vie for diminishing common sources flowing downstream or being pumped from common aquifers. Whether it is climate refugees looking for dry land or water to drink, it would seem there are challenges ahead that only get bigger and harder to manage the longer climate change continues unchecked.

## Applying the Philosophy: Making Climate-Savvy Decisions

Making climate-savvy decisions means determining the vulnerability of your work, community, or culture to climate change, and taking action to limit vulnerability by reducing climate change or its effects. The best climate-savvy planning incorporates explicit consideration of vulnerability, adaptation, and mitigation together.

Consider how a manager of protected lands might think about climate-savvy planning (fig. 4.3). By considering the vulnerability of her goal or system in relation to

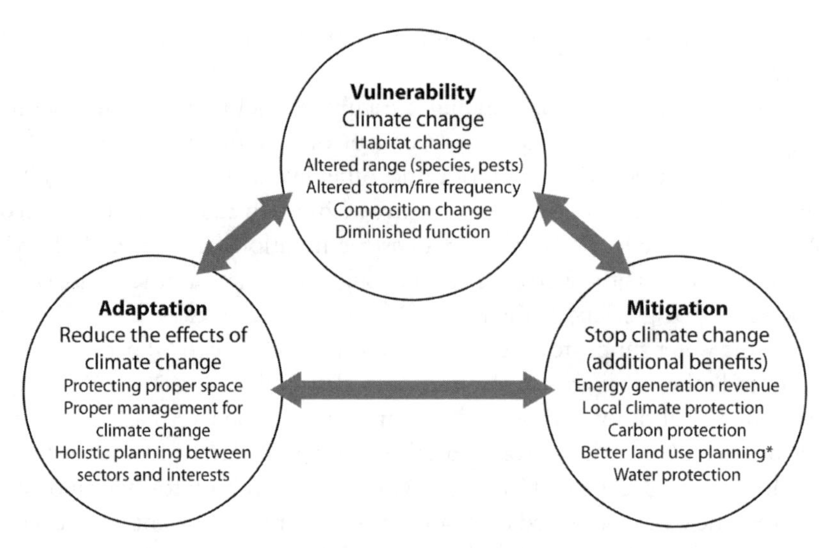


FIGURE 4.3 Interconnections among climate vulnerability, mitigation, and adaptation. All are important for climate-savvy conservation and resource management, and each can affect the others.

climate change she can focus on actions that will most benefit the goal or system, and ideally avoid wasting time and money on actions that climate change is likely to render useless or even damaging. Protected lands managers, for instance, will likely be concerned with habitat change, altered ranges of species of concern and pests, altered precipitation patterns, altered storm and fire frequency and intensity, changes in community composition, and diminished ecosystem function. Options for adapting land management to climate change could include focusing protection on areas likely to see the least change or to best withstand changes, adjusting management to respond to and minimize change, or collaborating with other landowners, regulators, and resource users on a holistic regional plan. Because the effectiveness and cost of adaptation depends in part on our success at limiting climate change, managers should consider mitigation as well, reducing their operation's greenhouse gas emissions by decreasing energy use, using onsite solar, wind, or biomass installations to meet some of their energy needs, engaging in land use practices that increase carbon storage and minimize carbon loss, or even allowing some larger-scale wind or solar power generation on their land if the location is appropriate. Some actions address mitigation, adaptation, and effective conservation and management all in one: working to improve regional zoning and limit sprawl decreases greenhouse gas emissions associated with longer transport times, increases landscape connectivity that supports plants' and animals' natural responses to climate change, and limits pollution and habitat loss.

There is not a simple panacea, nor are there right or wrong answers. Rather, there are a host of options that can reduce vulnerabilities and risks, each with different advantages and disadvantages for any given set of circumstances. The process of making a climate-savvy decision requires creativity and evaluation of numerous options within a broader, holistic context to determine which choices give the best likelihood of achieving the intended goal while reducing the likelihood of negative consequences in other places or sectors.

This type of decision-making is perhaps more straightforward when it comes to mitigation. You identify your goal, then evaluate a suite of options for meeting that goal in terms of how they do or do not reduce the risk of contributing to further climate change. This allows you to make a choice that results in lower greenhouse gas emissions while still achieving your goal.

For instance, when thinking about power generation, you can evaluate a range of options (coal, gas, solar, and so on) against the goal of generating more electricity while reducing greenhouse gas emissions (table 4.2). There is also the option of reconsidering your goal altogether. Is the stated goal really your ultimate goal? In this case, the question is whether the goal is additional electricity generation, or if it is instead to have sufficient energy to support a particular level of development and well-being. In the latter case, energy conservation may be a cheaper, more effective approach than increased generation.

To extend this thinking to individual decisions such as transportation, consider a trip across town to deliver a box (table 4.3). Here your goal is to get the box across town, with the added intent to minimize greenhouse gas emissions. Considerations include how far the trip is, the weight of the package, and what condition it needs to arrive in, as well as short- versus long-term goals (rapid delivery versus limiting climate change). You can drive a car, take the bus, ride a bike, walk, or do nothing. Which solution best meets your goals?

This same approach can be applied to adapting conservation and management to climate change. Here you blend the traditional goal (protect a species, place, community, or resource) with the added goal of decreasing the vulnerability of that goal to climate change. We illustrate the process for protecting the hydrology of the Yellowstone River in table 4.4.

Focusing on longer-term, underlying goals helps avoid undermining that goal's success for short-term benefit. For example, the recent increase in potential funding opportunities associated with climate change has many people hoping to get more funds by reframing their work in terms of climate change. Unfortunately, simply reframing your work as a climate solution without considering the vulnerability of your tactics to climate change can result in inability to meet your overarching goals over the long term.

One increasingly popular way to raise funds or obtain land for forest protection is marketing carbon credits, or offsets, for carbon held in the soils and trees of existing forests. From a forest conservation perspective, this is a great short-term idea. You get money to manage, maintain, or buy land and protect the flora and fauna within. You avoid carbon emissions from forest destruction or degradation. However, if only the mitigation corner of the triangle in figure 4.3 is considered, you miss the vulnerability of the forest itself to climate change. It may be possible to save the land on which the

**TABLE 4.2** Assessment of options for generating electricity while minimizing greenhouse gas emissions

| *Goal:* | Generate more electricity | | | | | |
|---|---|---|---|---|---|---|
| *Climate Change Risk Reduction:* | Do this with minimal greenhouse gas emissions | | | | | |
| *Options:* | Coal/Gas | Nuclear | Hydro | Wind | Solar | Alternative thinking |
| Pros | High energy output per unit<br><br>Currently cheap | Low emissions per unit energy<br><br>Potential for substantial energy generation | Low emissions per unit energy<br><br>Substantial energy generation | No emissions per unit energy | No emissions per unit energy | Is there an alternative to generating more energy that would also help us avoid emissions but still do the things that need to be done?<br><br>Options:<br>*energy conservation |
| Cons | High emissions per unit energy<br><br>Extraction damages the environment | Radioactive waste<br><br>Security<br><br>Vulnerable to climate change (reduced availability of cooling water)<br><br>Extraction is costly and damaging | Methane emissions from reservoirs<br><br>Habitat destruction<br><br>Vulnerable to climate change (reduced river volume) | Poor siting can cause wildlife impacts<br><br>Footprint (large wind farms only)<br><br>Requires storage or part of mixed grid | Materials for solar collectors expensive and extraction and processing damages the environment<br><br>Footprint (large solar arrays only)<br><br>Requires storage or part of mixed grid | *increased energy efficiency<br><br>*reduced demand |

**TABLE 4.3** Assessment of options for delivering a box across town while minimizing greenhouse gas emissions

*Goal:* Deliver a box across town

*Climate Change Risk Reduction:* Do this with minimal greenhouse gas emissions

| *Options:* | Do nothing | Drive a car | Ride the bus | Bike/walk | Alternative thinking |
|---|---|---|---|---|---|
| Pros | No transport carbon emissions | Can easily transport the box | Less personal carbon | No transport carbon emissions | Why do we want to get the box across town? Is the box itself needed, or just something in the box? Can the contents of the box reach the destination another way?<br>Say the contents of the box is a recipe book, and it is being transported because a specific recipe is needed.<br><br>Options:<br>*scan the recipe<br>*read the recipe over the phone |
| Cons | Box does not get there | Carbon emissions; no exercise | Need to carry box to and from bus | Need a way to carry box | |
| Results | Goal not met | Box delivered but not climate-smart (good short-term success, poor long-term success) | Goal met, more climate-savvy | Goal met, more climate-savvy | |

forest is located but lose the forest to climate change, as was described for much of the Amazon basin in chapters 2 and 3. This also means that there is little long-term reduction in carbon emissions, regardless of management, if the biota that held the carbon disappear due to climate change. Both the forest and the global climate lose. There are other reasons why basing forest conservation arguments heavily on carbon storage or sequestration could backfire. Some models suggest that some forest systems are expected to release more carbon than they take up as the climate warms. Would this mean that those who protected the forest and received funds for the stored carbon are accountable for these emissions? Other models suggest that the dark boreal forests absorb more heat than they reflect and cause a net warming of the global climate, even when their carbon storage role is taken into account (Bala et al. 2007). Thus timber

**TABLE 4.4** Assessment of options for protecting the hydrology of the Yellowstone River ecosystem in the face of climate change

| | | | | |
|---|---|---|---|---|
| *Goal:* | Protect the hydrology of the Yellowstone River ecosystem | | | |
| *Vulnerability to climate change:* | Region is warming and drying, resulting in less water in the ecosystem, higher water temperatures and altered timing of peak flow | | | |
| *Options:* | Do nothing different | Change extractive demands | Reintroduce beavers | Use snow fences |
| Pros | Easy to do<br>No immediate cost | Can make local development and agriculture more sustainable | "Makes mountains into sponges" | Easy to reverse |
| Cons | Nothing changes | Problems can occur upstream of extraction | Hard to reverse effects if they are adverse | Limited scope; requires sufficient snow, human maintenance and monitoring |
| Likelihood of success | Very low | Fair | Good | Good |

companies could argue for massive logging in boreal forests as a means to fight climate change, which is not a desirable outcome for conservation or sustainable resource use. There are many excellent reasons to protect or sustainably manage forest ecosystems, and it may be that promoting stewardship of nature rather than stewardship of carbon stores is a better long-term strategy.

## Final Thoughts

Adapting to climate change is not an endeavor to be taken on in isolation. Rather, it needs to be part of all of our decision-making and planning processes. Climate change is a reality that must be integrated into our most basic thinking, just as we have until now considered the historic climate, the attributes of what is present in a given location, the uses of the resources managed, or any other aspect of the places we work. All decisions, from species or forest conservation plans to urban development plans to water quality protection mechanisms to transportation planning, need to be climate-savvy because all sectors of our society and natural world are being affected by climate change. Climate change is our reality, and ignoring it will not make it go away.

# Chapter 5

# *Assessing Vulnerability to Climate Change*

Good judgment comes from experience, and experience comes from bad judgment.
—*Rita Mae Brown*

Thus far, this book has focused on why adapting to climate change is important, and on general principles for actually doing it. How can we turn all this into actual projects, or apply it to our own work? The first step is to assess the ways in which our goals and the species, places, and processes we care about are vulnerable to climate change. Understanding sources of vulnerability forms the basis for developing adaptation strategies; knowing the relative vulnerability of different species, places, or resources can help to prioritize where and how to focus our efforts.

Climate change vulnerability assessments can be as extensive as the global IPCC assessments that involve thousands of scientists and dozens of new climate models, or as simple as gathering a handful of people with the relevant expertise (including scientific, organizational, and management knowledge) in a room together to talk things through. They may focus on a single project, an organization, or the entire world, and may address just climate change or include it as part of a comprehensive vulnerability assessment process, recognizing climate change as one risk category out of many. For instance, a group considering whether a species should be listed as endangered would likely assess the vulnerability of that species to a range of stressors including pollutants, habitat fragmentation, and illicit harvest as well as to the effects of climate change.

However you do your climate vulnerability assessment, it is important to design and execute it in ways that maximize its relevance to and use by actual projects. Take the time to think about who will use the assessment results and how they will use them. If the target audience is people or organizations who make strategic decisions about

**BOX 5.1 ASKING THE RIGHT QUESTION**

Many organizations approach climate change vulnerability assessments as though they were asking, "What are all the possible effects of climate change, and what can we do about them?" This is not generally the most productive approach. Not only is it overwhelming, but unless your organizational mission is to save the world from climate change, it is not even appropriate. A better way to start might be to ask the question "What do we do, and how can we adjust that to account for climate change and its effects?"

resource allocation, they may want an assessment that identifies which places or species are most and least vulnerable. If the intended audience is people who develop recovery plans for particular species, they may prefer an assessment that describes sources of vulnerability for those species. There are a handful of best practices that apply across the board—using a suite of climate projections rather than just one, for instance—but other elements of vulnerability assessments must be tailored to the goals and objectives of those doing and using the assessment, the ecological and sociopolitical context of the assessment, and available skill, time, and money. This chapter reviews some basic concepts and approaches, and criteria to consider in deciding how to move forward.

## Components of Vulnerability

Vulnerability to climate change has three basic components: exposure to change, sensitivity to change, and capacity to adjust or adapt to change. Bruce Stein of the National Wildlife Federation uses an analogy with sunburns to explain. In the case of sunburns, exposure refers to the amount of solar ultraviolet radiation that hits your skin. Cave-dwellers have low exposure, and people sunbathing on tropical beaches at noon have high exposure. Sensitivity refers to how easily you burn. Pale-skinned people who burn after just a short time in the sun have high sensitivity, while people with plenty of melanin who can stay in the sun all day without burning have low sensitivity. Adaptive capacity refers to the ability of potential sunburn victims to effectively minimize their sunburn risk, for instance by using parasols or changing when they go outside. People whose jobs require them to be outside in the middle of the day would have less adaptive capacity than those with more control over when they are outside, and people with access to sunscreen would have more adaptive capacity than those without.

To reduce vulnerability to climate change, then, we can reduce the rate and extent of change and its negative effects (exposure and sensitivity), and enhance the ability of species or systems to recover or cope (adaptive capacity). Although climate change is a global phenomenon, vulnerability assessments and adaptation plans should be based on exposure, sensitivity, and adaptive capacity as manifest in the particular systems, places, or species under consideration.

## Start with the Basics

Vulnerability assessments should begin with an examination of the mission, vision, goals, and objectives of the group doing the assessment, and of the assessment itself (fig. 5.1). If the assessment takes place within a preexisting organizational context such as a governmental agency, wildlife refuge, or corporation, the mission, vision, goals, and objectives may already exist. If the assessment is being undertaken by a newly formed group of individuals or representatives of multiple organizations, the group as a whole should reach clarity on the overall mission of the assessment before proceeding. A community group with an overarching mission of maintaining profitable fisheries for the next century would want different information in a different form than if its mission were to maintain the social and economic health of a community that is currently built around fishing.

The next step is to assess the vulnerability of the goals and objectives themselves to climate change, and to determine whether they need to be adjusted in response to those vulnerabilities. As an example, consider the following goal for a hypothetical national park: *To restore the coastal freshwater wetland to its condition prior to European settlement so that it will once again support the diverse suite of bird, fish, amphibian, and plant species that historically called it home.*

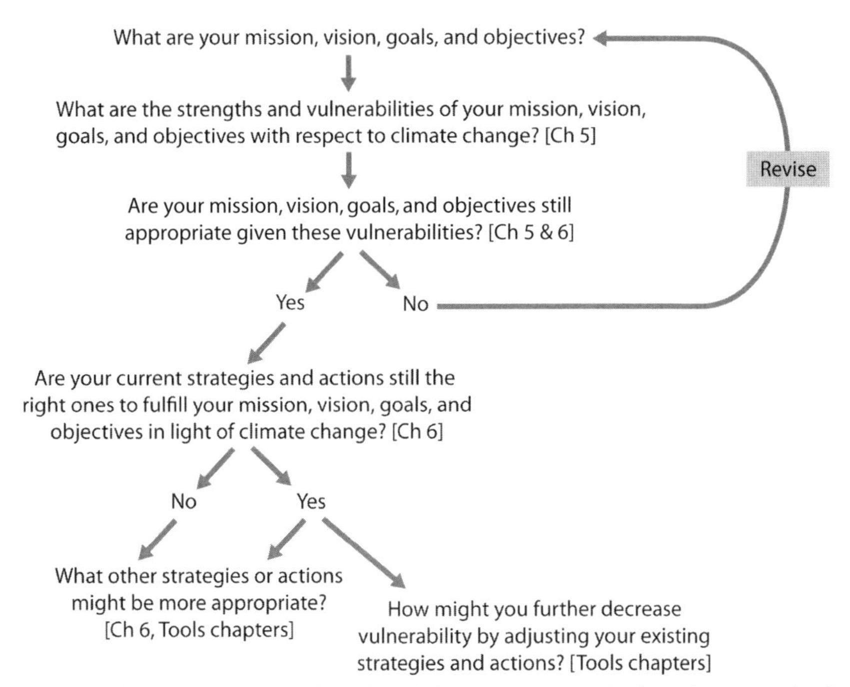


FIGURE 5.1 The flow of deliberation for vulnerability assessment and adaptation. Keep the focus on your organization or project's overall mission, vision, and objectives rather than on some abstract concept of climate change adaptation.

This goal is potentially quite vulnerable to climate change. The climate that supported the wetland and its inhabitants in centuries past has almost certainly changed due to global climate change and changes in land use in the surrounding area or watershed. Furthermore, sea-level rise will cause saltwater to gradually (or suddenly) intrude into the marsh, changing it from fresh to brackish. Maintaining the same suite of species or even the marsh itself without extensive and ongoing intervention may no longer be possible. Park managers should consider whether to stick with the original goal, knowing they may be committing themselves to expensive and ongoing efforts (this may be desirable if the marsh is economically, socially, and culturally important), or to change their goal to one less tied to past conditions. A less climate-vulnerable revision might be: *To restore the wetland in a way that allows its continued support of diverse and abundant plant and animal species as it shifts from fresh to brackish during the next fifty years*. Rather than focusing on resisting change, this goal supports the realignment of the ecosystem as climate and sea level change.

Our hypothetical national park might have goals that focus on public awareness as well as on the ecosystems and species themselves, such as: *Park visitors observe and understand the unique flora and fauna that live on or near glaciers*. Clearly, as glaciers disappear, it will no longer be possible for visitors to observe organisms living on or near glaciers. A less climate-vulnerable goal might be: *Park visitors understand the role that glaciers play in physically shaping the park and the plant and animal communities throughout the entire watershed*. Even if the glaciers disappear entirely, visitors will still be able to see their influence in shaping the park. Indeed, educational programs can be developed to illustrate how plant and animal communities move and change as glaciers shrink and disappear, raising awareness about how climatic change affects ecological systems.

Most vulnerability assessments involve multiple partners or stakeholders, and the goals of the organizations, communities, or individuals undertaking the assessment may not be the same—they may even conflict—and the design of the vulnerability assessment must reflect this. There are a variety of motivations for doing vulnerability assessments, including:

- To prioritize species, locations, or programs across a broad region to target future efforts and funds
- To develop detailed management plans for a particular site or population
- To provide legally defensible support for whatever decisions an agency makes as a result
- To use as a mechanism for raising funds
- To underline the need for reducing greenhouse gas emissions by highlighting the possible consequences of not doing so
- To underline issues of social injustice

Clearly, different objectives require different types of assessments. To meet the first objective, a relatively coarse-scale assessment would be sufficient, while the second would require a much more detailed assessment. The first three objectives are

**BOX 5.2 WHO TO PROTECT?**

The importance of goal clarification is illustrated by a story from the South Pacific. An international aid organization had funded the construction of a sturdy seawall in front of a village to protect it from the combined effects of sea-level rise and storm surge, and from that perspective the project was a success. A village member later commented, however, that they would much rather have protected the graveyard, because while living people were able to move their houses, the dead were pretty much stuck where they were. Thus from the local perspective, the project had focused on a problem of lesser importance, leaving a major issue unaddressed.

best met by an assessment that clearly incorporates degrees of certainty and provides a solidly realistic output. The final three, in contrast, could be effectively met with more generalized inputs and assumptions. Explicitly addressing differing goals and motivations creates a more transparent process, increases the chances that stakeholder needs will be met, and provides an opportunity to scale participant expectations to available resources.

## The Context of the Assessment

Having established overall goals and objectives, the next step is to consider the sociopolitical and biophysical context of the vulnerability assessment.

On the biophysical side of things, climate change should be viewed within the context of regional climatology, hydrology, and ecology. For instance, species and communities that are used to greater climate variability throughout the year may be less sensitive to change than those that are adapted to a more uniform climate. Tewksbury and coauthors (2008) suggest that for this reason, tropical species may be more vulnerable to climate change than temperate species, even though the magnitude of change is expected to be smaller in the tropics. Characterizing climatic variability in space as well as time in the target region can help to reveal the availability of microhabitats (valleys that remain cool, north-facing slopes, and so on) that may provide refuge as climate change progresses. A solid understanding of the basic climatology of the region of interest highlights the particular climatic elements that are most important. Likewise, an understanding of how water normally moves through the system in question and where it comes from is essential to informing the scope of the assessment. How much of the water comes from snowmelt? From coastal fog generated by ocean upwelling? From underground springs? The answers to such questions may lead to expanding the assessment beyond the originally intended boundaries.

The sociopolitical context influences all elements of vulnerability (exposure, sensitivity, and adaptive capacity) as well as how best to carry out the assessment. Demographic characteristics such as projected changes in population and resource use can significantly influence the climate vulnerability of species and ecosystems. Cultural values, existing rules and regulations, the political or social structures through which those rules and regulations are generated and enforced, and the effectiveness of the enforcement influence both climate vulnerability and the range of possible adaptation options. For instance, climate change will play out much differently in a community with poor regulatory enforcement that allows excessive deforestation or water use than it will in a community with strong environmental protection laws and enforcement. For species threatened with imminent extinction by pollution or invasive species or communities ravaged by armed conflict, it may make sense to do only a rough assessment to determine whether climate change poses enough of a threat to warrant diverting attention from more immediate threats.

## Top-Down versus Bottom-Up Approaches

The terms top-down and bottom-up refer both to the degree of public engagement—from a handful of experts gathering and analyzing information that they then share with stakeholders, to a process where all stakeholders are engaged as equals from the beginning—and to the focus of the assessment, whether it begins with a broad look at potential changes or starts first with particular local concerns and questions.

To the extent that they arise from existing management, regulatory, or planning concerns, bottom-up assessments can provide a solid basis for putting climate-savvy planning, management, or conservation into action. They start by identifying elements that are essential to the structure or function of the system of interest—anything from an ecosystem to a watershed to a human community—regardless of vulnerability (fig. 5.2a). Essential elements might include areas key to the reproductive success of

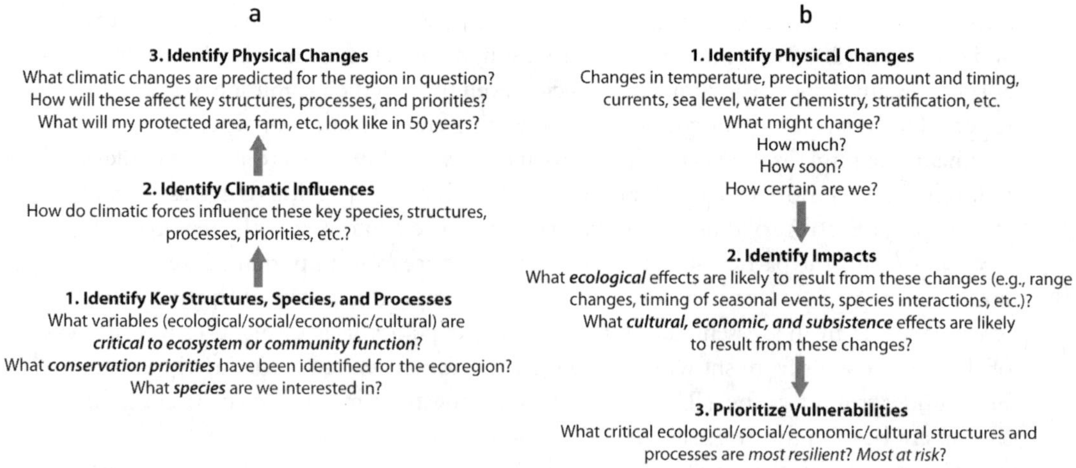

a

**3. Identify Physical Changes**
What climatic changes are predicted for the region in question?
How will these affect key structures, processes, and priorities?
What will my protected area, farm, etc. look like in 50 years?

**2. Identify Climatic Influences**
How do climatic forces influence these key species, structures, processes, priorities, etc.?

**1. Identify Key Structures, Species, and Processes**
What variables (ecological/social/economic/cultural) are ***critical to ecosystem or community function***?
What ***conservation priorities*** have been identified for the ecoregion?
What ***species*** are we interested in?

b

**1. Identify Physical Changes**
Changes in temperature, precipitation amount and timing, currents, sea level, water chemistry, stratification, etc.
What might change?
How much?
How soon?
How certain are we?

**2. Identify Impacts**
What ***ecological*** effects are likely to result from these changes (e.g., range changes, timing of seasonal events, species interactions, etc.)?
What ***cultural, economic, and subsistence*** effects are likely to result from these changes?

**3. Prioritize Vulnerabilities**
What critical ecological/social/economic/cultural structures and processes are *most resilient*? *Most at risk*?

FIGURE 5.2 The conceptual flow of bottom-up (a) and top-down (b) vulnerability assessments.

species of interest, particular soil chemistry, or an aquifer that provides a reliable source of water for a village.

The second step in a bottom-up assessment is characterizing climatic influences on the priority elements identified in the first step. The focus is not on climate change per se, but on the influence of climatic factors in general. It is important here to identify not just general influences (total annual rainfall, average monthly maximum and minimum temperatures, and so on) but also specific aspects of the climate system that may serve as triggers or thresholds for species or processes (e.g., box 5.3). The timing of the first frost or the frequency with which a particular temperature threshold is passed may be more important than monthly average temperature. Without understanding the role of climate today, it is difficult to know what effects changes in climate might have.

The third step involves an exploration of possible climatic changes, targeted toward those climatic variables identified as being particularly important in the previous step. Gather all appropriate and relevant data, not just the easily available data. If little or no regional information on projected changes in the important variables is available, at least this process highlights gaps in data and modeling that need to be filled.

A top-down approach (fig. 5.2b) typically starts with a broad view of potential changes, and works down to potential effects of these changes on species, places, or processes. Projected changes, both climatic and ecological, are usually more generic than those in a bottom-up assessment because they are not driven by as focused a set of concerns. Top-down approaches are good for characterizing the relative vulnerability of a range of places, species, or systems at once, and for exploring when and where it might be important to incorporate climate change into planning, management, or conservation. For example, an organization whose goal is to prevent extinctions might do

### BOX 5.3 HIDDEN BY AVERAGES

When Anton Seimon of the Wildlife Conservation Society set out to do a vulnerability assessment in Africa's Albertine Rift, he discovered that there was very little published climatic information for the region. Most descriptions of its climate were fairly vague statements about rainfall seasonality and ENSO-related variability. Before investing in downscaled regional climate models, the WCS team realized it needed a better understanding of the region's current climatology. Using a range of conventional and unconventional data sources, the team revealed a complex picture of regional climate patterns. In some locations, the two rainy seasons were broken up by periods of intense rainfall followed by brief periods of relatively low rainfall, a pattern that had been previously masked by the use of monthly precipitation averages. Such patterns could provide phenological triggers for flowering plants and migratory species. Had the team jumped straight into climate modeling, they would have completely overlooked what may be a critical feature of the regional climate.

a top-down assessment to identify species at particular risk from climate change. An organization looking to purchase land for conservation might use an assessment to identify places least vulnerable to climate change.

In the end, many groups combine elements of top-down and bottom-up assessments, often integrating information from a range of disciplines as well. A group whose initial aim was to identify and subsequently focus their organizational effort on the most resilient areas may find after a top-down assessment that a site it cares deeply about is highly vulnerable. Rather than abandon the site, the group may reassess its prioritization scheme and use the information from the top-down assessment for a quick bottom-up look at the key vulnerabilities of the site and to develop adaptation options. Natural scientists focused on the vulnerability of a lake ecosystem in a heavily agricultural area may decide to include social and agricultural scientists in their assessment process to effectively capture possible risks or benefits related to the farming communities' response to climate change.

## Types of Information

Most vulnerability assessments use some combination of ecological, climatic, physiochemical, and socioeconomic information. While it is useful to explore the wide variety of information available, it is also important to clarify what information is most important for the assessment at hand. Some of the desired information may be missing. For instance, downscaled climate models providing projections at fine spatial resolution are immensely useful in some situations, while in others the availability of data or the understanding of current climatology may be too limited to make such modeling worthwhile. This does not mean that vulnerability assessment and adaptation planning cannot proceed. Indeed, it may help practitioners and other stakeholders to fully acknowledge the uncertainty inherent in any planning process. Climate change simply adds another layer of uncertainty.

### *Physical, Chemical, and Climatic Changes and Their Effects*

Although the popular media often discusses climate change in simplistic terms of changes in average temperature or rainfall, the effects are much more diverse and often subtle (fig. 5.3), including:

- Changes in the timing, amount, and type of precipitation
- Changes in air, soil, or water temperature regimes, including average and extreme values as well as the frequency and magnitude of extreme events
- Changes in wind and current patterns
- Changes in stratification, turnover, and upwelling in lakes and oceans
- Changes in the frequency, intensity, or characteristics of storms
- Changes in sea level

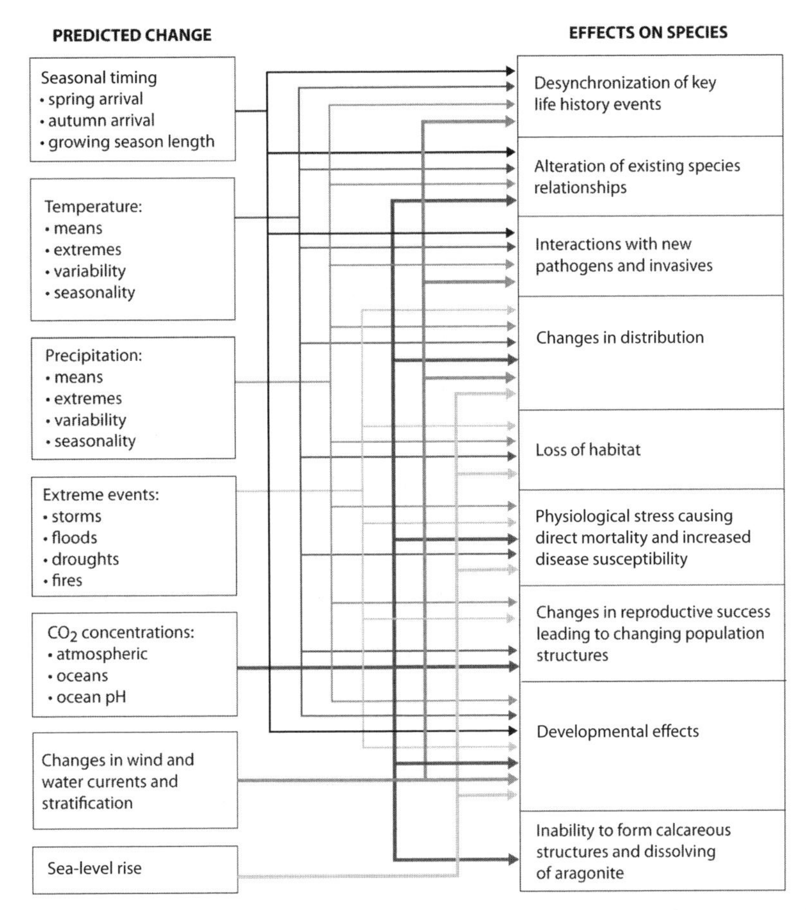


FIGURE 5.3 Likely effects of physical and chemical changes tied to greenhouse gas emissions on species. Any given change has multiple interacting effects, and any given species is affected by multiple interacting changes. After Foden et al. 2008.

- Changes in hydrological regimes, such as the magnitude, timing, and duration of high- and low-water periods
- Changes in water chemistry, such as ocean acidification

The wide range of physical and chemical changes can lead to an equally wide array of biological effects (fig. 5.3), but as always, being clear about goals can help to target your efforts. If the goal is to prioritize species for conservation or management attention, you might focus on variables relating to reproduction and survival of a range of species types. If the goal is to create a robust forest harvest strategy, you might focus on variables related to forest health. Some categories of effects to consider include:

- Changes in timing of seasonal events, such as earlier bloom, migration, or leaf-out, and the potential for mismatches when two linked events do not change

their timing at the same rate; for example, a bird nests at the same time, but the primary food item for its chicks appears earlier
- Species range expansion, contraction, or shifts to higher latitude or altitude, or into favorable microhabitats
- Changes in species interactions such as competition or predation may lead to changes in range or local population densities (including local extirpation)
- Changes in critical habitat availability
- Changes in ecosystem services such as pollination or flood protection
- Changes in pest or disease outbreak frequency and severity, or the arrival of new pests and diseases
- Effects on culturally important areas

For migratory species or those whose habitat use changes significantly throughout their life history, it is important to consider possible effects in all habitats (e.g., breeding *and* wintering grounds; freshwater larval habitat *and* terrestrial adult habitat), and how changes in one habitat might affect species' use of other habitats. For instance, improved feeding opportunities for snow geese on their winter feeding grounds have caused a rapid population increase that has led to widespread deterioration of summer breeding grounds (Jefferies and Drent 2006).

### *Interactions between Climate Change and Other Stressors*

As discussed previously, climate change can exacerbate and be exacerbated by numerous other stressors. Including other stressors in your climate vulnerability assessment (fig. 5.4) thus provides a more comprehensive assessment and decreases the chance of unpleasant surprises. While many climate change vulnerability assessments have taken the simplified approach illustrated in figure 5.2, integrated approaches are becoming increasingly common. Possible interactions include:

- Increasing conflict over water use in areas where drought frequency and intensity increases
- Increased toxicity of particular chemicals due to changes in temperature or salinity, or effects of pollutants on temperature tolerance of plants and animals
- Increased runoff of pollutants from land into aquatic environments, particularly the possibility of periodic inputs of high intensity when a heavy rainfall event occurs following prolonged drought
- Shifting patterns of human resource use—for example, fishers becoming farmers as lakes dry up, or farmers becoming fishermen as drought makes farming too unreliable
- Habitat fragmentation preventing range shifts or preventing gene flow that might facilitate evolutionary adaptation
- Deforestation further increasing drying and severity of flooding and erosion

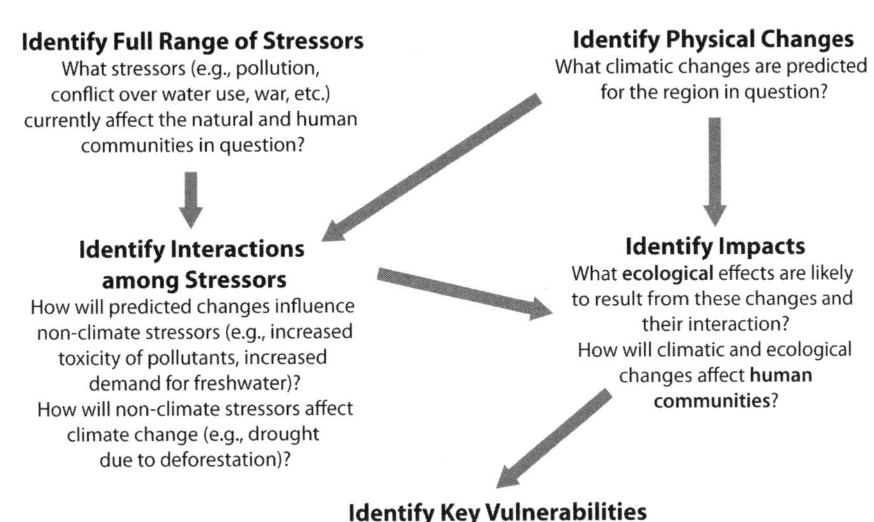


FIGURE 5.4 Flow of an integrated climate vulnerability assessment. These assessments address the full range of stressors, including climate change, as well as interactions among these stressors. The critical vulnerabilities for any particular species or system may result from climate change, other stressors, or the combination thereof.

### *Setting Priorities*

Because there are too many possible foci and too many climatic and nonclimatic variables to do a detailed analysis of everything, you will have to prioritize what receives the most attention in your analysis. Which structures and processes are *most likely* to be affected? Which structures and processes are *most important* to community or ecosystem function? The answer to these and similar questions, not the availability of data or models, should determine where the greatest effort is expended. Importance may be evaluated relative to the species, place, or system itself, or relative to the possible effects on those systems. Thus considerations might include:

- Ecological importance: keystone species, ecosystem engineers, functional groups (decomposers, primary producers, consumers), sources of limiting factors such as water or nutrients
- Economic importance: direct sources of income—for instance through tourism or the production of marketable resources—or providers of ecosystem services such as flood control, food supply, or water availability
- Cultural importance: sources of food, medicine, or traditional materials—species and places central to spiritual rituals or other traditions
- Immediacy of threat: sooner versus later
- Persistence or reversibility of effects

- Potential for adaptation to compensate for effect
- Distribution of impacts: national versus local, income groups affected, and so on

## Vulnerability or Sensitivity Indexes

Sensitivity or vulnerability indicators can help to organize information gathered during a vulnerability assessment, and to develop and monitor adaptation options. Indicators are essentially those factors that play a role in determining sensitivity or vulnerability to climate change, and once a list of indicators is compiled one can assign scores to species, habitats, or other assessment targets for each factor. In some cases, factors are weighted to reflect their relative contribution to overall vulnerability. Thus an absolute dependence on a vulnerable habitat type, such as sea ice, might be given more weight than other factors. Factors that influence community and population vulnerability include inherent biological traits of organisms (e.g., rapid evolutionary rate, flexible resource requirements, length of larval period), physical characteristics of the location (e.g., the existence of microhabitats, relatively pristine condition), or the social and economic setting (e.g., level of political stability, influence of extractive industries).

Indicators may be kept as sets of distinct scores, but are often converted to standardized scores that are then combined into a single index (e.g., Laidre et al. 2008). While these aggregate indexes have the advantage of appearing relatively simple, they often mask important information about sources of vulnerability. A species with an average aggregate score could be moderately vulnerable to all factors, or it could be extremely vulnerable to one factor but not particularly vulnerable to others. Combining individual scores into a single index might be useful if the goal is to rank relative vulnerability or sensitivity to climate change, while individual factor-by-factor scores are more useful for identifying possible interventions for reducing vulnerability.

Indicators have been proposed for a variety of taxa, and efforts are under way to develop a range of indicators that could be used to assess and monitor change in the southern ocean. The U.S. Geological Survey, for example, has developed a coastal vulnerability index for the United States that uses data on tidal range, wave height, coastal slope, shoreline change, geomorphology, and historical rate of relative sea-level rise to quantify shoreline vulnerability to the effects of sea-level rise.

## Sources of Information

Potentially useful sources of information for assessing vulnerability include observation and measurement, comparison with analogous situations, experimentation, models, and expert judgment and local knowledge. Combining multiple sources of information may compensate for the biases or weaknesses of any source alone. For instance, laboratory experiments on the temperature tolerance of a particular species may suggest potential tolerance thresholds, and field observations can help to illuminate whether

## BOX 5.4 GLOBAL SPECIES VULNERABILITY TRAITS

Wendy Foden of the International Union for the Conservation of Nature (IUCN) is spearheading an effort to develop globally useful tools to identify potential effects of climate change on a range of taxonomic groups. The project grows out of the recognition that IUCN's existing Species Red List Categories and Criteria were developed before climate change was widely perceived as a threat to biodiversity. During a four-day workshop a diverse group of experts described traits that made species more vulnerable to extinction in general, then identified those traits most important for determining sensitivity to climate change. The workshop focused on birds, amphibians, and warm-water reef-building coral. Workshop participants identified five categories of traits that increased species vulnerability to climate change:

- Specialized habitat and or microhabitat requirements
- Narrow environmental tolerances or thresholds that are likely to be exceeded
- Dependence on specific environmental triggers or cues that are likely to be disrupted by climate change
- Dependence on interspecific interactions that are likely to be disrupted by climate change
- Poor ability to disperse or colonize a new, more suitable location

wild populations respond similarly to laboratory populations. It may be that certain factors in the wild lead to different temperature tolerances than are found in the laboratory.

*Observation and monitoring* are at the heart of understanding climate change, its effects, and the effectiveness of any efforts we may make to reduce vulnerability to it. In addition to standardized monitoring programs established as part of research or management plans, citizen-science efforts such as the annual Audubon Society Christmas Bird Count or even personal journals can provide useful information. There is growing interest in repeating historic censuses and comparing past and present results. Although often described as being as exciting as watching paint dry, a well-designed and well-enacted monitoring program provides both the data and the familiarity with a particular system needed for effective climate vulnerability assessment and adaptation.

The *empirical analogue* approach uses past and present phenomena as a guide to what future changes might bring. For instance, the response of species and ecosystems to periods of rapid warming in the paleontological record may provide some indication of possible responses to current change. Likewise, responses to annual or decadal climate cycles such as El Niño and the Pacific Decadal Oscillation provide insight into possible effects of short- and long-term climate change on landscapes and communities. The strength of this approach is that it provides integrated system responses across a wide range of scales. Its weakness is that responses to past change or shorter-term change may not be good indicators of responses to current long-term change. In particular, the existence of a wide range of confounding or exacerbating conditions such as habitat loss, pollution, and overharvest may dramatically reduce the ability of populations or species to respond successfully to climate change.

*Experimentation* is a useful tool for understanding the mechanisms underlying the response of species and communities to change, and an experimental mind-set is a key element of adaptive management (discussed further in chapter 16). Experiments can address everything from the physiological response of individuals to a particular stressor in the laboratory to the community- or ecosystem-level effects of a combination of stressors in the wild. The strength of this approach is that it explicitly tests the significance of particular stressors or processes for the system of interest. The weakness is that you must strike a balance between naturalness and the ability to manipulate the variables of interest.

*Models* can address a range of questions, from how the climate system might change to the effects of those changes on species, habitats, or community interactions. A vulnerability assessment can make use of many models, for instance by using climate models to explore how the climate system itself may change, and by using biophysical, economic, or integrated system models to investigate possible effects of those changes on ecological and human systems. Models can be developed to address all kinds of variables, providing a sense for possible future scenarios and relevant considerations for conservation and management planning. The reliability of model outputs is highly dependent on the assumptions used to make the model and the availability of data for validating them. Even the best models produce only projections of what the future might be like, not guaranteed predictions. There is a degree of irreducible uncertainty about the future that must be accepted and built into planning and management.

*Expert judgment and traditional knowledge* provide qualitative information that is difficult to capture in models or experiments. Most critically, they help to characterize uncertainty and identify critical factors that may have been omitted or that cannot be quantified. People or communities who rely directly on hunting, fishing, or other forms of wild harvest often have a nuanced understanding of the species and habitats they rely on for food, and long-standing communities may be able to provide knowledge of past climate patterns that goes back generations.

## Getting the Right People on Board

Once you have identified the goal of the assessment and the approach you are going to take, the next step is to identify the people needed to make the assessment successful and useful. Consider not just people with scientific knowledge or technical skills, but people who represent the values and perspectives of the relevant communities, who have extensive traditional ecological knowledge, and whose buy-in and assistance will be important for disseminating assessment results and enacting adaptation plans. While outside expertise or technical experts can provide essential skills and knowledge, relying on them too extensively can weaken an assessment by inappropriately discounting local knowledge and common sense. We discuss who to engage and how to engage them more thoroughly in the next chapter.

## Final Thoughts

The practice of formal vulnerability assessment came into being well before widespread concern about climate change, and has been used to address a range of hazards. Climate change vulnerability assessments are simply a new evolution in this older field. Climate-focused vulnerability assessments can be useful for building knowledge and capacity around the implications of climate change for resource management, but climate change can also be addressed as one more element of a broader vulnerability assessment. They are not rocket science, although they do require a thoughtful approach, and they need not be carried out in a vacuum.

# Chapter 6

## *Developing Strategies to Reduce Vulnerability*

> You can't always get what you want. But if you try, sometimes, you get what you need.
>
> —*Mick Jagger and Keith Richards*

Having decided to engage in climate-aware planning and action, how should we go about it? As with vulnerability assessments, there is a rich literature categorizing approaches to adaptation planning and a multiplicity of categorization schemes. The key is to keep the focus on overarching goals and objectives (fig. 6.1), and to build or use an approach that works best within your ecological, climatic, and sociopolitical context. If your current goal is not to solve all the problems arising from climate change, that does not have to be your future goal either. As Adamcik and coauthors (2004) commented about developing refuge strategies:

> It is important not to choose strategies without objectives, develop objectives without goals, or establish goals without first articulating a vision. Otherwise, two common errors may result: (1) You may develop goals to justify existing management programs and then create a vision that incorporates them; and/or (2) You may choose strategies you already are using (e.g., burning, grazing, partnerships), and then develop objectives to justify them. The result may be a plan that validates existing management practices, instead of one that objectively considers alternative actions and then directs effort toward achieving refuge purpose(s) and vision. Articulating a vision based upon refuge purpose(s) and other mandates will allow you to identify existing programs that may need to be refocused or eliminated.

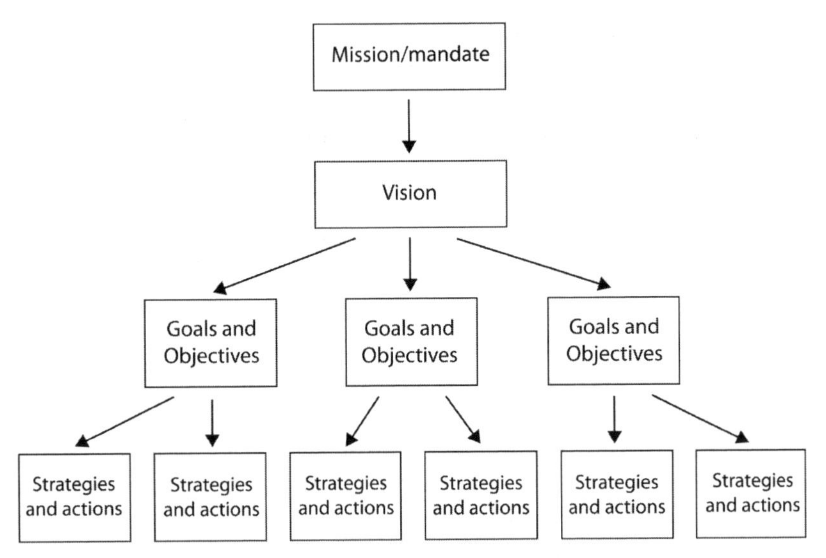


FIGURE 6.1 Hierarchical relationship of project or organizational mission, vision, goals and objectives, and action. Where possible, climate vulnerability assessments and adaptation planning should take place within existing organizational frameworks, not separate from them.

In this chapter, we discuss general principles for effective adaptation, commonly cited obstacles to making adaptation happen, and approaches for putting adaptation into action on the ground.

## General Principles

The principles outlined below are not unique to climate change adaptation. They grew out of other fields, but address the core issues that are so important to successfully adapting to climate change: accounting for uncertainty and human decision-making processes, appropriately engaging the people who will provide the information and support you need, and creating systems that allow you to maximize learning while doing.

### *Resistance and Resilience*

Two key concepts in vulnerability reduction are resistance (the ability to weather disturbance with little change) and resilience (the ability to bounce back after disturbance). A coral reef that experiences stressful temperatures without bleaching is resistant, while one that bleaches but rapidly recovers is resilient. In some cases we know enough to develop strategies specifically to increase resistance or resilience to climate change. Restoring healthy beaver populations or appropriate levels of grazing can make an area more resistant to drought, for instance (see chapter 11). In the absence of specific vulnerability information or targeted adaptation options, supporting general resistance and resilience by reducing overall stress on a system is a reasonable approach, although likely less effective than a more climate-informed approach. It is also worth noting that exposure to stressful conditions over many generations may in some cases *increase* the resistance or resilience of affected populations as natural selection weeds out less robust individuals.

There is sometimes a trade-off between resistance and resilience. An emergency command center should be designed and built to remain functional throughout anticipated disasters (resistance), for instance by having walls strong enough to withstand hurricanes and being on high enough ground that flooding will not be an issue. In contrast, less essential infrastructure could be designed for resilience through ease of repair, mobility, or other approaches. Assateague Island National Seashore in Maryland, located on a dynamic barrier island subject to strong seasonal storms, has taken the latter approach with its mobile bathhouse project. Each lightweight bathhouse has solar-powered showers, passive solar vault toilets, and other design features allowing it to be easily moved out of harm's way during storms or relocated permanently as the shoreline shifts. The seashore also paves roads and parking lots with crushed clamshells rather than asphalt to facilitate repair and minimize the amount of hard infrastructure interfering with natural processes.

### *Robust Decision Making*

The climate change community has focused heavily on reducing uncertainty around both projected climatic change and ecological responses to that change. While the goal of reducing uncertainty is laudable, we will never be rid of all or even most uncertainty. In particular, we will never know with certainty what future anthropogenic greenhouse gas emissions will be, making it impossible to predict the future climate with full certainty. Other important sources of difficult-to-resolve uncertainty include natural variability in climate and ecological systems and the trajectory of feedback loops such as methane release from melting permafrost. Thus we would do well to shift from asking what exactly the future will bring to asking what management decisions we can make now that give us the best chance of achieving our goals across a range of plausible futures.

This viewpoint engenders a shift in approaches to planning. Currently, a common strategy is to identify the single most successful outcome for an assumed future and to plan toward reaching that single best outcome. This framework works well if uncertainty is low, but may increase the likelihood of a negative outcome if uncertainty is high—that is, there is no guarantee that the assumed future or anything like it will come to pass, or that one's actions will have the desired result. A robust decision-making framework focuses on maximizing the probability of *some* successful outcome across a range of plausible futures, although we may not know at this point exactly which outcome will be possible. The key is to explicitly consider a range of possible actions and strategies and to map how each would play out over time in multiple future scenarios. Rather than committing fully to any single strategy, you can select near-term actions that leave an acceptable set of future options open across all scenarios. As the future unfolds, you can continue to map out options and select the most robust choices.

### *Accounting for Human Behavior and Values*

Why do people or institutions fail to take action, make choices that seem to go against their own interests, or behave in what appear to be irrational ways? Scientists often act as though the problem is lack of awareness or knowledge, and simply providing more information will get people to take action or behave rationally. As elucidated by extensive research on human decision making, however, the truth is more complex. In our efforts to make climate adaptation a reality, we need to come to grips with several realities.

On an individual level, one common reason people fail to act is that they perceive themselves as having little or no control over a problem that seems hugely daunting. A strong perception of risk is often correlated with what is termed avoidant maladaptation, such as wishful thinking or fatalism, which inhibits even the intention to take action (Grothmann and Patt 2005). The conservation community's heavy focus on the dire and particular threats posed by climate change (the "climageddon" approach) coupled with fairly general admonishments to "act now" may in fact motivate apathy rather than action.

More broadly, scientific or factual information is only one element of any given decision-making process, and people simply do not make decisions in a consistent or

#### BOX 6.1 PROTECTING THE STRONG VERSUS THE WEAK

Having identified which populations or locations are more vulnerable or more robust, where should we focus conservation efforts? Some contend that protection should focus on the most robust populations or places because others may not survive in the face of climate change regardless of what we do. Others feel that protection should focus on the most vulnerable, since they will clearly not survive without human intervention. A third group argues for focusing on places or populations of intermediate vulnerability based on the belief that the most vulnerable will not survive in a changing climate no matter what we do, and the least vulnerable are likely to survive even without our help. Yet another group argues that prioritization should be based primarily on a place or species' emotional, spiritual, cultural, economic, or other importance regardless of climate vulnerability. Models suggest that the relative effectiveness of these approaches is highly contextual (Game et al. 2008), but there is little empirical evidence. In the end, the choice must be made by individuals or organizations on the basis of their risk tolerance, goals, and the status of the systems in which they work.

rational way. Some of these nonrational elements of decision making have to do with how an issue is framed—people are much more likely to accept a 25 percent chance of success than a 75 percent chance of failure—but others touch on deeply held belief systems that people bring to the decisions they make. Most of us have some values that are so important to us that they trump other interests, including financial ones. Centering arguments for adaptation around values or beliefs that are not universally shared by stakeholders—that humans have caused climate change, for instance, or that big government is bad—is unlikely to lead to success. Particularly when it comes to asking people or institutions to try new things, it is generally more effective to focus on needs or objectives rather than particular positions.

Finally, if we want to influence policy or management (or human behavior in general), we have to engage with the reality of how the people we want to influence do their work. Policymakers have hundreds of competing demands and do not have time for abstruse discussions of the subtleties of climate models. Managers do not typically read the peer-reviewed scientific literature as a matter of course, making that literature a less effective means of getting ideas into action. From an adaptation perspective, we need to expand climate change research beyond climatic and ecological phenomena to include the effectiveness or lack thereof of existing and suggested management options, and to disseminate our findings in venues that reach managers and practitioners as well as academics. The assumption so far has been that once we have enough information or the right information about climate change and its effects, we will easily be able to translate that knowledge into management or policy. Unfortunately, creating actionable information is not the same as creating action.

### *Stakeholder Engagement*

Engaging or informing a range of individuals and organizations in planning adaptation strategies is not always necessary or appropriate, but it is sometimes essential and often enriching. Stakeholders can provide data or other information that increases the robustness and usefulness of the adaptation planning process and the plans themselves, particularly when it comes to information that is not easily available online or in the published literature. They can also assist with the long-term monitoring and feedback essential for evaluating the success of adaptation plans once they are implemented. This is particularly important when adaptive management is part of the strategy. Further, adaptive management may require more stakeholder buy-in and flexibility than other strategies because managers may need to change their approach more frequently. In cases where adaptation will require stakeholders to change the way they think, such as engaging in planned retreat from sea-level rise rather than armoring shorelines and building right on the beach, stakeholder engagement can help to build support for a new way of doing things. The extent to which stakeholders are engaged and empowered can vary from simply informing them of the plans to including them as full partners at every step of the process (fig. 6.2). As always, decisions about who to engage at what level should reflect the goals and the context of the exercise at hand. If a reserve manager and those who use the reserve (e.g., hunters, birders, hikers) lead a vulnerability assessment and adaptation planning process, the results will likely have greater support, compliance, and longevity. An externally controlled assessment in which those stakeholders are not engaged may run more smoothly and be backed by a higher-level mandate. It is essential to be clear and up front about any constraints on the process or outcomes, such as legal, regulatory, or policy mandates. This limits the likelihood that stakeholders will end up feeling ignored or betrayed when their suggestions are not implemented. Providing the opportunity for input that is ignored can create more disempowerment than not providing the opportunity for input in the first place.

The most basic element of engagement is to be respectful of stakeholder needs, values, and previous efforts at adaptation. Take the time to find out what has been done and what was successful or surprising before developing a new plan, and identify essential elements of how community decisions and opinions are formed. In some cultures presenting a community with a completed plan, even a draft plan, effectively shuts down real input since rules of behavior prevent extensive criticism. Communities in

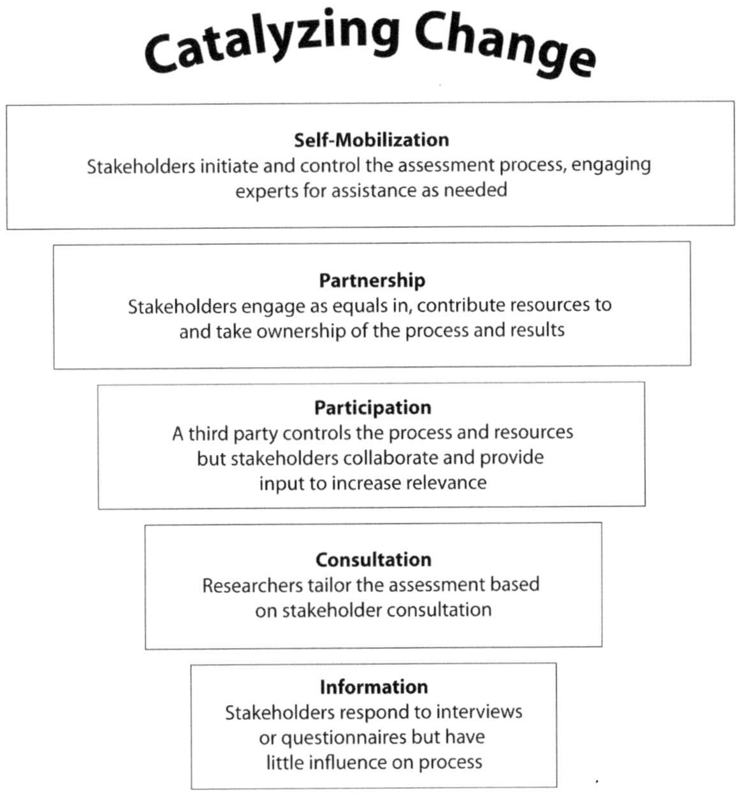

Catalyzing Change

**Self-Mobilization**
Stakeholders initiate and control the assessment process, engaging experts for assistance as needed

**Partnership**
Stakeholders engage as equals in, contribute resources to and take ownership of the process and results

**Participation**
A third party controls the process and resources but stakeholders collaborate and provide input to increase relevance

**Consultation**
Researchers tailor the assessment based on stakeholder consultation

**Information**
Stakeholders respond to interviews or questionnaires but have little influence on process

FIGURE 6.2 Range of stakeholder engagement options. Stakeholders are those whose actions and knowledge may affect or be affected by the planning process and its outputs. After Pretty et al. 1995.

particularly vulnerable areas such as small island states or the high Arctic may have been approached by multiple groups of outsiders looking to do adaptation projects. Failure by many of these groups to communicate with each other or to maintain a long-term commitment to the communities in question can lead to a distrust of unknown groups or individuals seeking to offer "help."

Lack of communication combined with insensitivity to local culture has derailed many a project, for instance the World Health Organization's (WHO) initiative to combat malaria by releasing male mosquitoes that rendered the females with which they mated sterile. With little advance communication a WHO van emblazoned with the organization's symbol, which includes a serpent wrapped around a staff, appeared in a village in India to begin field trials. Villagers despised and feared snakes, so the van itself aroused skepticism. Feelings did not improve when several foreigners emerged from the van and released clouds of mosquitoes. The second time the van appeared in the village, locals chased the WHO officials back into their van and out of the village. The third time, they burned the van.

As with vulnerability assessments, getting the appropriate set of people involved at the appropriate level in project planning and implementation can make the difference between success and failure. It is helpful if leaders of adaptation efforts are members of the community they are trying to influence, for instance if local efforts are led by locals or international efforts are led by international groups. Categories of stakeholders to consider engaging at various stages include those who will use the adaptation plans or be asked to change their behavior as a result of the plans; decision makers; opinion leaders; and those with relevant knowledge and experience. British Columbia's Integrated Resource Planning Committee (1993) outlines a series of steps to ensure that the scope of potential participants is adequately identified:

1. Create an initial list of organizations, interest groups, and individuals who may wish to be involved in the process or whose buy-in may contribute to project success or failure;
2. Meet with representatives of these groups separately in informal, low-key settings that are familiar to the people with whom you are meeting;
3. Explain clearly the principles of adaptation and the goals of the project with which you are asking them to engage, and ask about previous related efforts;
4. Emphasize the importance of public participation, and that you are asking them to decide among a range of options for engagement, both in terms of the level of involvement and the mechanism;
5. Ask group members to express their interests or concerns, and request the selection of a group representative to participate in an initial joint meeting of all the groups; and
6. Ask these interested parties if they know of others who should be involved in the process.

Because climate change adaptation requires a holistic approach as well as creative new ideas, think broadly about the types of individuals and organizations with whom to engage. Consider possible cross-sectoral challenges and how to minimize the likelihood of different sectors working at cross-purposes. A classic example of different sectors negating the usefulness of each other's work comes from Vietnam, where international conservation, livelihood, and disaster agencies replanted and restored mangrove forests, some of which were cut down within just a few years by federal agencies promoting aquaculture.

### BOX 6.2 COMMUNITY-BASED ADAPTATION IN FIJI

On Kabara island along the southeastern boundary of Fiji's waters, people depend heavily on natural resources not just for livelihoods but for day-to-day subsistence. Working with the World Wide Fund for Nature (WWF) South Pacific Program, villagers engaged in a two-day, community-wide project known as Climate Witness (World Wide Fund for Nature–South Pacific Program 2005). Villagers collectively mapped the location of natural resources they used, and created a calendar of resource use throughout the year. They then discussed climatic influences on these resources and their availability, what changes in climate or resources they had already seen, and what future changes might reasonably be expected. On the second day, community members focused on values, sharing individual values first and then agreeing on values that most mattered to them as a community as well as their collective vision for the future. This was followed by a root cause analysis to understand factors contributing to undesirable changes identified on day one, and then by a number of processes geared toward developing a broad array of adaptation options. The final stage of the project was to develop a concrete Community Adaptation Plan (CAP) that reflected the resource needs, community values, and climate-related threats identified in the previous two days. Kabara village has carried out many elements of their CAP, including getting international grant money for water storage tanks to address their concern over increased salinization of drinking water.

The success of this project rests largely on the fact that the community had ownership of the vulnerability assessment and adaptation planning process. WWF staff provided information as needed and guided discussions, but villagers took responsibility for fully exploring and identifying vulnerabilities to climate change, then developing their own ideas for solutions. Local knowledge and insight provided results that could not have been developed externally.

A second key element is that adaptation planning placed as much emphasis on community values and priorities as it did on the physical aspects of vulnerability. After identifying the risks posed by climate change to natural resources, community members focused on what mattered to them as a community. This meant that they internalized not just how climate change might affect their natural resources but what this might mean for the social structure of the community as a whole. This increased the commitment of the community to the adaptation plan that was created; not only did they have ownership of the process, but their core values as a community were woven into the fabric of the final plan.

### *Monitoring Success*

Although long-term monitoring is not widely seen as "sexy" and as a result can be difficult to fund, a well-designed and well-executed monitoring program can address a plethora of needs. In addition to measuring the success of adaptation efforts, it can facilitate deeper understanding of climatic and ecological systems in the project area, provide data needed to validate models of climate change or its effects, measure effects of climate change, and build stakeholder capacity in and support for climate change adaptation. Monitoring is also an essential component of adaptive management, as discussed in chapter 16. In addition to questions related to project-specific efforts, central questions for developing a monitoring protocol include:

- *Do existing monitoring systems or programs provide the information needed to detect trends in the variables associated with key management decisions?* This information might include extremes in temperature, changes in salinity in estuaries, reproductive timing of key populations, or changes in soil or water pH. Even in areas with existing programs, the monitoring locations, timing, and parameters may not provide needed information.
- *Which environmental variables are most sensitive to climate change in the system of interest?* At the level of individual species, consider a range of life history stages, differences in habitat or resource use throughout the year, and the physical, chemical, and biotic variables most important to the status of individuals and populations. At the community or ecosystem level, consider which populations are known or assumed to be most or least sensitive to change, and which physical, chemical, and biotic variables are likely to change most rapidly.
- *How is the information gathered likely to be used by policymakers, managers, and other stakeholders to inform conservation or resource management decisions and behavior?* The way in which data are gathered, analyzed, and presented should reflect end-user needs. Thus a project geared toward implementing and improving human community adaptation efforts may do best with a low-tech community monitoring program that builds support and awareness, while a project geared toward building climate change into the recovery plan for an endangered species might require a monitoring program designed to feed into statistically rigorous analysis and modeling.

## Overcoming Obstacles

Many obstacles, real and perceived, prevent people from taking on the challenge of adapting their work to climate change. Some obstacles, such as limited funding or personnel time, are external, whereas others, such as apathy, fear, misinformation, or lack of empowerment, are internal. Below we discuss some common reasons people give for why they are not yet able to adapt their work to climate change, and possible approaches to addressing them. Not all obstacles can be overcome, but many are not as great as they seem.

### *Competing Concerns*

Climate change is just one of many challenges facing natural resource professionals. Clearly, the myriad concerns that have occupied us for decades—pollution, overharvest, habitat loss, and so on—are still there. Climate change may not be the biggest immediate threat, and managers may feel that if they do not resolve the immediate threats now, their conservation or management targets may not be around long enough for climate change to matter. An important point here is that adapting your work to climate change does not mean focusing your work on climate change. Pollution or habitat loss may indeed be the biggest and most immediate threat to the species or habitat of concern, but climate change can influence which strategies will be most effective at addressing those threats or the magnitude of the threats themselves. It may also be that there are important options for adapting to climate change that are relatively simple or affordable now but will become prohibitively expensive or complex in the next ten or twenty years. For instance, preventing development in what will become critical habitat as a result of climate change is much easier than asking people to move once they are already living there. Thus incorporating climate thinking into your work can both improve your success at addressing immediate nonclimate stressors as well as set the stage for greater long-term success with your overall conservation or management goals.

### *Lack of Resources*

Lack of resources is an almost universal complaint among natural resource management and conservation professionals. Yet many individuals and organizations manage to achieve great things with little money, and we know of at least one case where an organization retuned some of its money to a funder—they felt that having too much money was leading to pressure from interest groups wanting in on the wealth, making it difficult to focus on the core goals of the project. Lack of resources can certainly be a real constraint, but it need not be a barrier to all adaptation action.

Take the time to focus once again on your organizational goals, and consider what it is you need to do, what specific activities you or your organization would undertake if more resources were available. Resist the urge to say simply "we would do more monitoring" or "we would do a better vulnerability assessment." Specifically what would you monitor? What specific vulnerability information are you missing, and how would having more information affect the ultimate course of action? What can you do with the resources you have now?

One option for tackling adaptation that can sometimes have lower incremental costs is to make certain that new and existing projects are in fact robust to climate change, for instance by including climate change in the environmental impact assessments performed prior to project approval or in the overall vulnerability assessments performed for threatened or endangered species. This approach reduces the risk that resources will be invested in projects whose success is highly vulnerable to climate change, or in actions that increase the vulnerability of other projects or sectors to climate change. There are a number of examples of municipalities taking this sort of approach, for instance the inclusion of changes in sea level and sea ice in the design and construction of the Confederation Bridge linking Prince Edward Island to mainland Canada.

### *Lack of Information*

Some people feel they lack the information or expertise necessary to tackle climate change adaptation. While more information may be useful, we can take solid adaptation action using readily available general climate projections, existing ecological knowledge, good planning processes, and techniques for managing under uncertainty. As Meffe and Viederman (1995) point out for conservation in general, lack of knowledge is rarely the real limiting factor for taking action. Waiting to act until uncertainties are resolved often prevents taking action until effects of climate change become apparent. Such post-facto adaptation is more costly and less effective than proactive, anticipatory adaptation. Furthermore, more data or modeling may not change the suite of actions that are possible or effective in response to a range of future conditions. We must shift our focus from reducing uncertainty to managing for it.

### *Lack of Tools and Guidelines*

A related issue is that individuals and organizations may feel that they "do not know how" to do vulnerability assessments or adaptation planning, citing a lack of available tools and guidelines as a reason for inaction. Yet hundreds of such guidelines and tools exist (e.g., UNFCCC 2008), and in fact such overwhelming choice itself can be paralyzing. A growing number of professional societies and government agencies are developing support systems to help those in their field incorporate climate change into their work, which should make finding the right tools for your particular situation easier.

While technical knowledge and expert guidance may be useful, there is nothing magical about adapting your work to climate change. At its most basic, doing climate-savvy work just means building a new set of variables into existing planning and management structures. We have found that when presented with examples of adaptation projects on the ground, many people are astonished by how straightforward the process can be and how much can be accomplished without intensive technical input or fine-scale data.

### *Institutional Barriers*

Legislation and regulation are often described as being too inflexible for the dynamic planning needed to address the uncertainty associated with climate change. For instance, many federal agencies have fixed five-, ten- or even twenty-year planning cycles, and making changes during each cycle is difficult. Likewise, organizations are often risk-averse and constrained by rules, routines, procedures, and precedents. At the leadership level, acceptance of climate change and its implications is often slow, and it can be difficult for individuals within the organization to bring about the change that is needed. Yet there are often ways to work within existing structures to effectively incorporate climate change and related uncertainties. While not directly focused on climate change, the Department of the Interior's guidelines for doing adaptive management in the context of federal mandates and regulations provide an example of overcoming this sort of perceived barrier (Williams et al. 2009). Presenting concrete examples of action by others can also help to assuage apprehensions. Adaptive governance is discussed further in chapter 16.

### *Lack of Empowerment or Political Will*

A related issue is lack of empowerment or political will. Perceived dichotomies between "experts" and stakeholders turn stakeholders into people who wait to be told the "right" way to do things or to receive "valid" information. In this model, stakeholders and practitioners may feel that they can do nothing until they hear from the experts. In reality, there must be dialogue and a truly participatory approach to developing and sharing knowledge. Science that aims to be not just relevant but actually *used* must be driven at least in part by an understanding of what managers and practitioners do, what tools they have, and what decisions are pending on what timescale (Vogel et al. 2007). Adaptation efforts need scientific, management, and governance information, not scientific information alone, to succeed.

## Taking Action

At the most general level, adaptation strategies are built around the three components of vulnerability: exposure, sensitivity, and adaptive capacity. Having explored the sources of vulnerability for a particular species or system, we can brainstorm options for reducing exposure or sensitivity and increasing adaptive capacity. It is important to distinguish between short-term coping and longer-term adaptation. The former will get you through an immediate crisis but does not necessarily provide a good foundation for longer-term vulnerability reduction. Both may be necessary at times, but ongoing coping is not the same as adaptation.

In general, incorporating climate change into existing processes is most likely to lead to action on the ground in the near term. Where planning processes or partnerships at the appropriate scale do not exist, new ones should be developed in pragmatic ways that account for the reality of the way the relevant partners function.

### *Reworking Existing Conservation Tools, Strategies, or Plans*

At a minimum, we should ask how climate change may influence the effectiveness of existing policies, tools, and management strategies. Will they remain effective, or are they vulnerable to climate change? Can they be adjusted to account for climate change? Will they lose relevance completely? As an example, consider the different types of equations managers may use in determining the maximum sustainable yield (and thus allowable harvest levels) in various fisheries. One approach is to calculate the maximum sustainable yield (MSY) based on the fishing effort and catch for a particular species over a number of years. If climate change alters the population dynamics of the species of interest, say by changing death rates, age to maturity, or the carrying capacity of the environment for that species, past data on yield and effort may no longer be relevant. In contrast, an approach that explicitly incorporates the species' population growth rates or the environmental carrying capacity for that species could adjust values for those variables to reflect known or projected effects of climate change. If climate change causes the rate of population growth or carrying capacity to fluctuate less predictably and beyond the current range of variability, all existing MSY models may become ineffective and a new approach could be needed.

Begin the process of developing adaptation strategies by considering the tools you currently use and the actions or decisions you currently have the authority or personnel to implement. Do you regulate harvest or pollutants? Make land decision purchases? If your agency makes opportunistic rather than strategic decisions about what land to purchase or protect, a tool to evaluate whether a particular acquisition is worthwhile given the reality of climate change may be more useful in the short term than a tool that supports strategic prioritization of lands to acquire. Over the medium or longer term, the latter may be useful to inform a handful of targeted acquisition efforts or to highlight areas in which personnel should be particularly alert to acquisition opportunities. If your organization or agency has a fairly flexible scope of action, you can think broadly about options for adaptation. If you have less flexibility, consider what can be accomplished using the tools at your disposal, then identify and work with others who have the needed resources or authority to address critical adaptation needs that you cannot.

There have been a range of efforts to incorporate climate change into existing plans, tools, and processes. A consortium in Nova Scotia created guidelines for building climate change into the environmental impact assessment process and tested those guidelines against existing projects (Bell et al. 2003), and a number of endangered species recovery plans explicitly include climate change (e.g., Povilitis and Suckling 2010). In the United States, there has been a concerted effort to develop and test guidance for incorporating climate change into state wildlife action plans (Association of Fish and Wildlife Agencies 2009).

### *Developing Targeted Adaptation Options*

Location- or target-specific plans can be developed in a number of ways. Where climate change effects have not been deeply explored before, it may be best to start with a fairly general climate vulnerability assessment for the selected community or location (fig. 6.3). Stakeholders then select a target or targets for action in light of organizational or

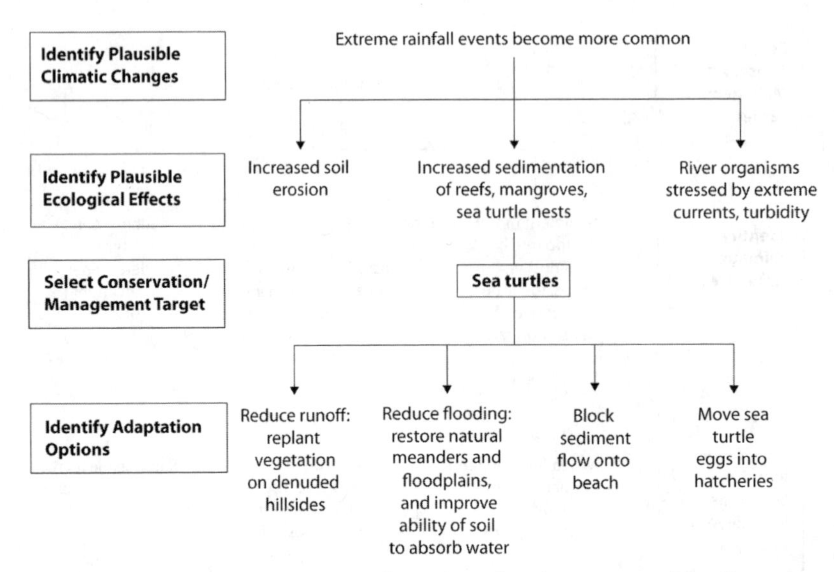


FIGURE 6.3 Schematic of a top-down adaptation planning process. The first two steps—identifying plausible changes and their effects—are a vulnerability assessment. The last two steps use information from the assessment to develop adaptation strategies. This schematic includes just one climatic change and three plausible effects; a real vulnerability assessment would clearly include more.

societal priorities and values as well as the results of the assessment. To explore adaptation options, stakeholders should consider the following questions:

- How can we reduce physical or chemical changes in our region? In the example illustrated in figure 6.3, there is little to be done locally or regionally other than reducing the rate and extent of global change.
- How can we reduce the negative effects of those changes, either in terms of ecosystem function or in terms of particular habitat types or taxa? Consider both more nature-based (restoring lost vegetation or connectivity) and more interventionist (sediment-blocking walls, putting eggs in hatcheries) approaches.

Take the time for creative brainstorming, letting the ideas flow without censorship. Once a number of ideas have been generated, evaluate them with respect to likely effectiveness, sustainability, possible negative effects, and other criteria (see box 6.3). In the example here, for instance, blocking sediment flow onto beaches would likely increase the rate at which the beach erodes, which in the long run would increase vulnerability to climate change for turtles nesting on that beach.

Alternatively, the adaptation planning process can begin with a focus on particular species or systems (fig. 6.4). Here, participants consider how climatic factors influence the assessment target, explore those elements of climate change that have the greatest

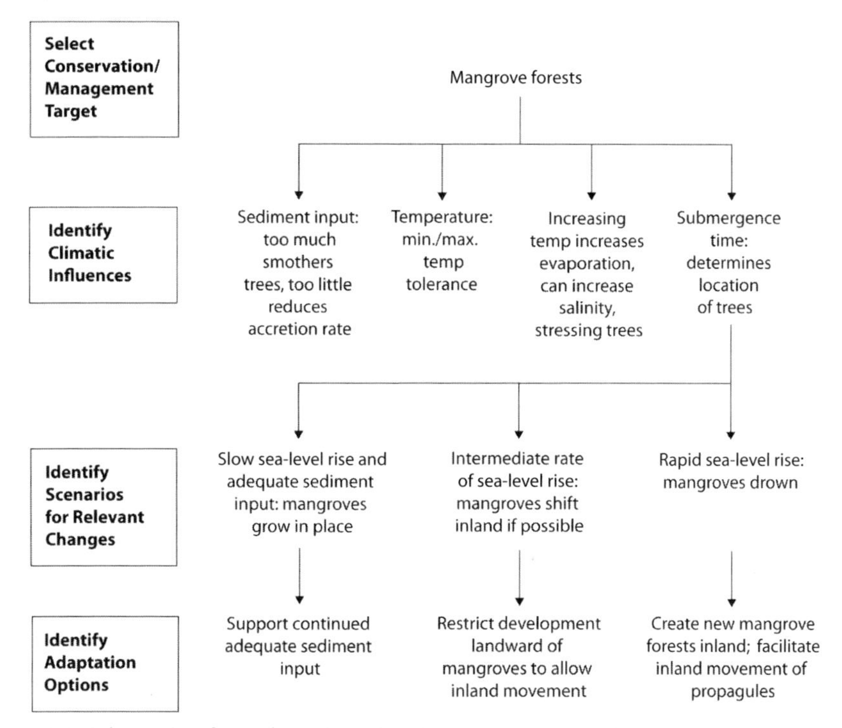


FIGURE 6.4 Schematic of an adaptation planning process in which the conservation or management target is known at the beginning. The second two steps are a vulnerability assessment for that target, followed by the development of adaptation options.

influence on the focal system, and brainstorm ways to reduce climatic changes or minimize negative effects on the focal system or species. In the case of mangrove forests illustrated in figure 6.4, we might consider three sea-level rise scenarios and explore adaptation options for each. Again, adaptation options should be evaluated critically before being acted on.

Clearly, figures 6.3 and 6.4 illustrate a simplified and partial view of adaptation planning; an actual vulnerability assessment and planning process would be more complex. In determining the sources of vulnerability on which to focus, rank the sources of vulnerability in terms of their contribution to total vulnerability and the ease with which you can address them, and combine this information to balance acting on easy targets with addressing critical vulnerabilities.

### *Regional Scale Planning*

International planning bodies have been developed for a number of purposes, including managing internationally shared resources (e.g., the International Baltic Sea Fishery Commission), managing or conserving migratory species (e.g., the North American Waterfowl Management Plan), or coordinating conservation for large landscapes (the

Central Albertine Rift Transboundary Protected Area Network). Likewise, cooperative planning bodies or processes have been developed within individual countries to increase the effectiveness of conservation or management efforts by coordinating diverse actors (e.g., Landscape Conservation Cooperatives in the United States). Such larger-scale coordination will become increasingly important as climate change alters the location, timing, and availability of species and resources, and can be made more robust by anticipating such changes. For instance, if populations of a commercially harvested species are likely to shrink in some countries and expand in others as a result of climate change, revenue-sharing or other agreements may be put in place before a crisis emerges.

Even absent established coordination processes, managers and regulators can communicate across jurisdictions about anticipated threats and challenges emerging in response to climate change. For instance, officials in one jurisdiction may warn those in adjacent jurisdictions when a noxious species appears poised to expand into new areas, giving them time to establish early response protocols. Likewise, they may provide advance advisory if commercially or recreationally harvested species appear to be shifting into new jurisdictions, and advise on effective management and regulation for those species.

Multijurisdictional planning and cooperation is likely to become increasingly contentious in cases where already limited resources become further diminished and one jurisdiction develops adaptation plans affecting resource availability in others, for instance building dams or other structures that reduce the supply of water or sediment to communities downstream. Organizations and agencies should consider and plan for such scenarios in developing adaptation options. Are there critical resources whose supply could be restricted by those in other jurisdictions? If so, what agreements or actions might be taken now to limit the possibility of such restrictions coming to pass, or having negative consequences if they do?

## *Evaluating Options*

Having come up with an exciting and creative suite of possibilities for adapting to climate change, how can we prioritize or decide among them? Existing decision-making processes provide a solid framework with which to start, since they presumably reflect organizational cultures and mores, although most will need some adjustment to become climate savvy. Box 6.3 lists a range of criteria by which we might evaluate adaptation options.

In addition to these considerations, a broader philosophical question is how interventionist to be. As previously discussed, adaptation strategies can range from a do-nothing approach that lets nature take its course to highly interventionist strategies such as covering glaciers with plastic or introducing species to areas beyond their current range (fig. 6.5). Neither end of the spectrum is inherently better, although organizations and individuals may have values or approaches to risk that bias them toward one end or the other. It is important to evaluate realistically whether the options under

### BOX 6.3 EVALUATING ADAPTATION OPTIONS

Taking adaptation action is good, but jumping into it without evaluating the options carefully across a range of criteria is not so good. The following considerations will help in the design, evaluation, and selection of adaptation options (adapted from de Bruin et al. 2009 and Titus 1990).

- Importance: What is at stake if you do nothing? Are there unique or critical resources whose vulnerability will be reduced?
- Urgency: What are the costs of delaying action, both in terms of what you might lose and in terms of what it would cost to implement later rather than now?
- No regrets and co-benefits: Do the benefits (including non-climate-related benefits) exceed the cost of implementation? Will there be significant beneficial outcomes even if the adaptation benefits do not pan out as expected?
- Economic efficiency: What are the expected benefits of this strategy relative to using the same resources elsewhere?
- Cost: How costly will the strategy be in terms of time, money, or other resources?
- Effect on climate change: Will the strategy increase the emission of greenhouse gases, or lead to undesirable changes in the local or regional climate?
- Performance under uncertainty: What is the strategy's likely performance across the range of plausible changes in climate for your region?
- Equity: Does the strategy benefit some people, places, or interests at the expense of others? Will this strategy have strong negative effects on any people, places, or interests?
- Institutional feasibility: Is the strategy possible given existing institutions, laws, and regulations? To what degree is the public likely to accept the strategy?
- Technical feasibility: Is the strategy technically possible to implement? Do we have or can we access the necessary tools and other resources?
- Consistency: Is the strategy consistent with existing national, state, community, or private values, goals, and policies?

consideration can achieve the desired goals, and the risks associated with each option. While maintaining the status quo often seems like a low-risk approach, it may be quite risky for species or systems for which the threats posed by climate change are clear and immediate. Conversely, the crisis mentality with which many approach the issue of climate change may create a temptation to experiment with more engineered or interventionist approaches. While there is logic to this approach in some cases (if a glacier or reef is clearly going to disappear in the next twenty years, wild and crazy ideas may be the only option), taking drastic measures when they are not yet necessary can be quite harmful.

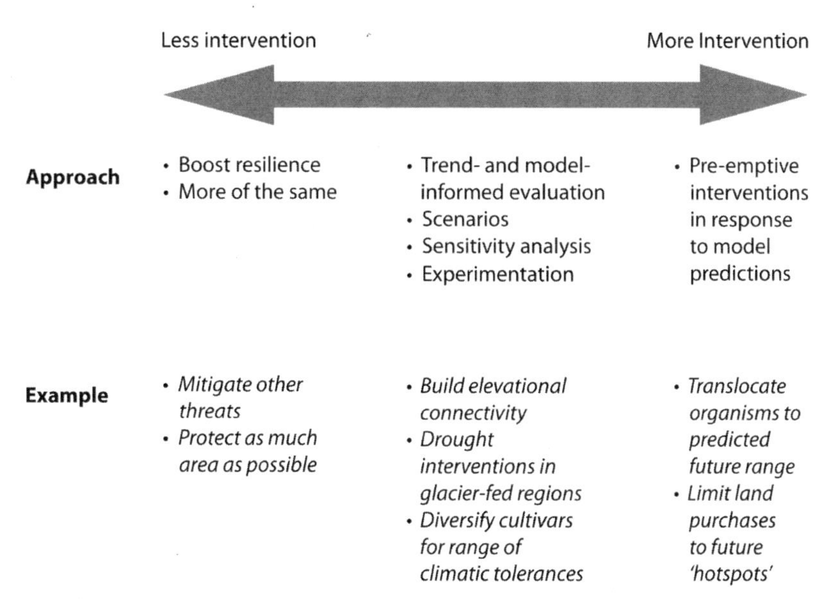


FIGURE 6.5 The range of adaptation options along an intervention continuum. For each level of intervention we include general approaches and an example. After Heller and Zavaleta 2009.

## Final Thoughts

The process of developing adaptation plans and strategies, if carried out appropriately, can itself help to decrease the vulnerability of the people and organizations engaged in that process. It does this by building awareness of and support for adaptation, spreading and integrating knowledge across a range of participants, and focusing people on what they can do rather than on what will happen to them. Because climate change adaptation by its very nature focuses on human action, it may at last move people beyond fatalism, apathy, or avoidance. The trick is to initiate the process.

# Chapter 7

# *Using Models and Technology*

> Models are like religion: you can have more than one, and you don't have to believe them.
>
> —*Daniel Pauly*

Models are a prominent element of climate change science and of efforts to adapt to climate change and its effects. Scientific projections of the future, such as statements about how much warmer it will get by what date or how a particular species' range will change as a result of climate change, are based on models. Models can help to develop further knowledge about a system or to address specific needs, such as developing management actions and policies that achieve conservation goals under a wide range of climate scenarios and ecosystem impacts. Even those of us who do not do modeling ourselves need to understand a bit about how they work and what they can and cannot do.

In the most basic sense, a model is simply a synthesis of current thinking and data about how a particular system works. Where systems are well understood, models may represent well-accepted ideas. Where processes are not well understood, the hypotheses illustrated by a model may be more controversial. Models may be conceptual or quantitative and can be expressed physically, graphically, verbally, mathematically, or in any way that facilitates communication, prediction, or hypothesis-testing. However they are expressed, they provide a framework within which to understand the world around us and make predictions about the future. A good model explains as many elements of the observations as possible yet remains simple enough to be tractable. In other words, models strike a balance between reality and utility.

Because past climate may no longer be a reliable predictor of the future, models are an essential element of exploring possible futures. Model output is *not* the same as hard data, and is not "true" in the same sense that actual observations are "true" (although the "truth" of observational data is affected by a range of factors such as instrumental or sampling bias). When using or evaluating models, consider how well they explain existing data (essentially "predicting" the present or the past—back- or hindcasting), how accurately they predict the future (although this is not possible to evaluate over long time horizons), and the degree to which they are consistent with other knowledge, such as basic laws of physics.

Here we present a *very* brief overview of modeling approaches that appear frequently in the climate change literature. Our goal is to convey both the value and the limitations of models. The former is important for understanding the appropriate use of models to elucidate aspects of ecosystem function and to inform conservation and management decisions, while the latter is important to prevent overconfidence in or misuse of model results.

## Climate Models

In order to make effective use of climate model output, we need to understand some essential information about climate models: what they are based on, how they are used to simulate past and project future climate, and what their primary strengths and weaknesses are.

### *What Is a Climate Model?*

Generally speaking, a climate model is a mathematical representation of our current understanding of the relationship among various factors controlling the climate system. At the core of climate models are basic physical laws of the universe, such as Newton's laws. In cases where the underlying physics of a phenomenon is not understood or where a physical process occurs at spatial scales smaller than can be resolved by the model, some statistical (mathematical) models are also embedded within these physics-based climate models. In some of these cases, the best information is purely empirical (observational as opposed to mechanistic), making statistical models the only reliable option.

Climate models may differ in terms of their internal structure (physics), how modelers define the relationships among the various internal parameters, and what initial conditions they use. The major differences among climate models generally relate to parameterization, or the way in which the relationships among variables are defined in a particular model. This comes into play for phenomena such as rainfall that cannot be modeled directly based on physical principles. For instance, precipitation is modeled not by using physics that directly control precipitation (which are not entirely understood) but by defining a particular relationship between precipitation and atmospheric variables such as temperature, humidity, and pressure.

## BOX 7.1 CLIMATE MODELING VOCABULARY

**AOGCM**: Atmosphere/Ocean General Circulation Model. Global, three-dimensional computer models of Earth's climate system, based on the physical, chemical, and biological laws that govern their component parts and interactions among them. Current AOGCMs incorporate complexities such as changes in land use or interactions among the biosphere, ocean, and atmosphere.

**Downscaling**: Generating projections at a higher resolution than provided by the grid cells of global climate models. May be *statistical* (based on mathematical relationships between large-scale inputs and fine-scale outputs) or *dynamical* (based on regional climate models; see RCM).

**Feedbacks**: Changes in the climate (real or simulated) wherein the initial response to some influence is either reduced (negative feedback) or enhanced (positive feedback) relative to the initial response without any further changes in input. A classic example of a positive feedback is the loss of snow and ice cover in the Arctic. Melting snow and ice usually expose a darker surface such as ocean water or land, which then absorbs more heat than the ice did. This increases the local warming, which melts more ice, and so on.

**Forcings**: External or internal factors that cause the climate (real or simulated) to change, for instance changes in solar radiation or greenhouse gases.

**Model**: A synthesis of hypotheses and the current state of knowledge (understanding) about how a particular system works

**Multi-model ensemble**: A group of model outputs created using a set of runs from different models. Averaging across multiple models brackets a suite of potential outcomes derived from a variety of approaches. Where most models produce similar results, we gain confidence that the process being modeled is relatively well understood and the results may thus be more reliable (although it may be the case that all the models are making the same error!).

**Perturbed physics ensemble**: A group of climate model outputs created by running the same model thousands of times with slight differences in the value of internal conditions and parameters each time, that is, the results of a sensitivity analysis for a particular model.

**RCM**: Regional Climate Model. Climate model covering a limited geographical area (such as a single state or region), at finer-scale spatial resolution than an AOGCM. They commonly use AOGCM outputs to provide boundary conditions (the broader climate setting) in order to simulate the response of regional climate to global trends.

**Simulation**: A single run of a model; usually refers to mathematical/computer models.

*Readers interested in more detailed discussion of climate models are directed to Randall et al. 2007.*

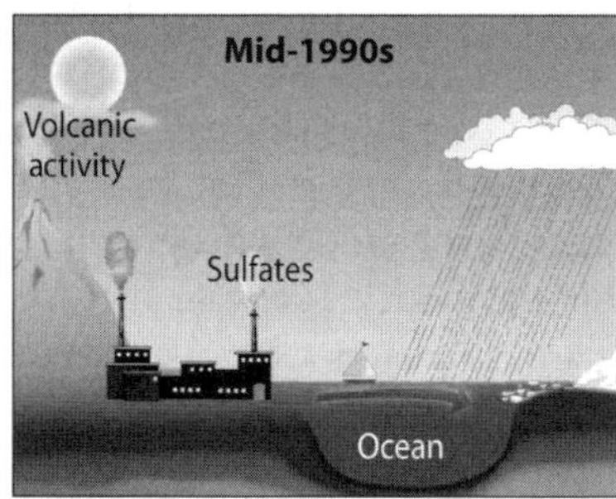


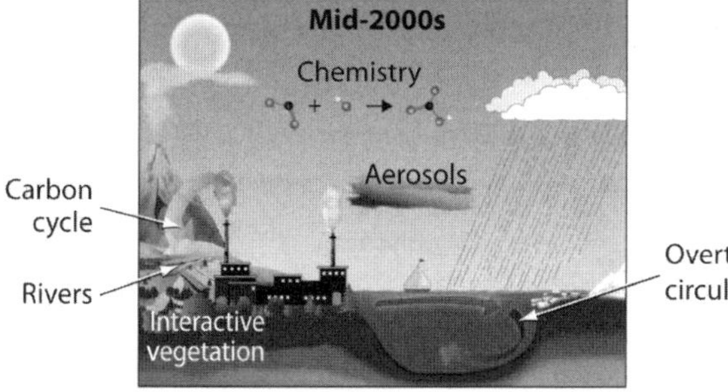


FIGURE 7.1 The evolution of climate model complexity over time. After fig. 1.2 in IPCC 2007a.

## *General Circulation Models*

Global climate projections are typically based on general circulation models (GCMs). These models have become increasingly sophisticated and realistic over time (fig. 7.1). Early GCMs included only atmospheric processes, essentially wind, radiation, and simplified clouds. In the mid-1990s modelers began including processes involving the upper ocean and sea ice, and used more sophisticated approaches to simulating clouds. This gave rise to the coupled ocean-atmosphere models (AOGCMs) that are now used for assessments such as the IPCC. Today, models include a wider range of factors, including aerosols, atmospheric chemistry, and the biosphere.

Century-long simulations by AOGCMs are run at a relatively coarse physical scale, meaning that these models can adequately simulate large-scale processes but become increasingly uncertain at smaller scales. Grid cells are typically around 100 to 300 kilometers on each side, meaning that features smaller than a single grid cell—many lakes, mountains, and islands, for instance—cannot be captured by the model.

### *Regional Climate Projections*

Assessing how climate change is likely to play out in any given region requires climate projections on a finer scale than that provided by AOGCMs. There are two general approaches to generating regional-scale climate projections: statistical and dynamical downscaling. All downscaling approaches address the effects of local topography, land use, climate patterns, and the like, but do so in different ways.

Dynamical downscaling relies on regional climate models (RCMs). Like GCMs, each RCM represents current thinking about the physical processes governing the climate system but this time at regional to local levels. Also like GCMs, they are physics-based models, but include regional physics. Each RCM must be run using a larger model such as a GCM (for future projections) or reanalysis (for past simulations) that provides regular updates to the boundary conditions, that is, that provides values for globally driven climatic conditions at the edges of the model's geographic area which then drive the regional climate changes within the model. The scale of RCM output is commonly on the order of 900 to 2,500 square kilometers (equivalent to squares that are 30 to 50 kilometers on a side), although grid cells may be as small as 5 kilometers on a side (fig. 7.2).

Statistical downscaling (SD), also known as empirical-statistical downscaling, uses mathematical estimates of the relationship between climate variables on a large scale

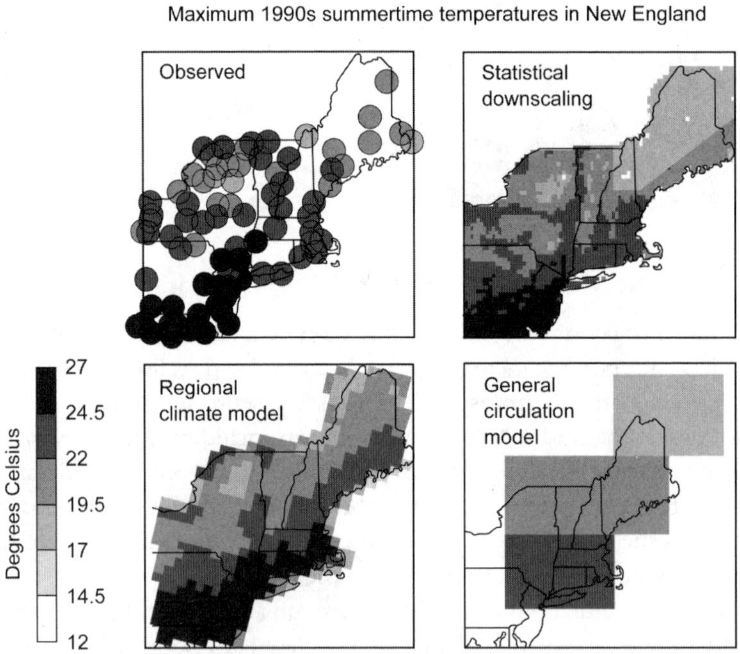

FIGURE 7.2 Comparison of observed and modeled temperatures in New England using three types of climate models. GCMs provide projections on a very coarse scale, while RCMs and statistical downscaling are better able to capture finer-scale projections. After Hayhoe et al. 2008.

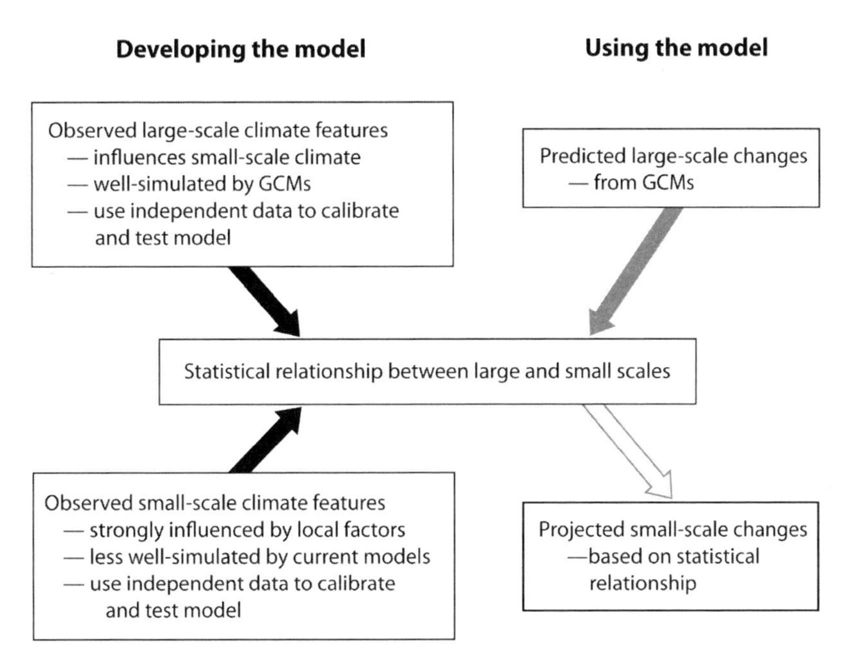


FIGURE 7.3 Schematic illustrating the development and use of a statistical downscaling climate model.

(e.g., daily temperatures over a larger region, such as New England) and observed climate variables on a smaller scale (e.g., daily temperatures for a particular city within the region) to generate downscaled climate projections (fig. 7.3). A variety of statistical approaches, including regression and correlation analysis, neural networks, or principal component analysis, can be used to link climate variables across scales. As with any statistical analysis, the method must be appropriate for the data set to which it is applied. For instance, a linear regression approach would not be appropriate in situations where data are intercorrelated. Also, there must be enough climate data available on the larger and smaller scales to both calibrate the statistical model (determine the correct relationships among the selected variables) and then test it against observed data that were not used in the calibration process. A central assumption of using SD to make climate projections is that the relationships between local and large-scale climate will remain roughly the same under future climate conditions.

Because RCMs make no assumptions about whether the relationship between small- and large-scale climate phenomena will remain constant, they have often been seen as more scientifically robust than SD methods. Regional climate models are also capable of simulating small-scale physical processes that can strongly affect the response of a given region or locality to climate change. On the other hand, RCMs take much more computing power, and because they must be independently developed and run for each region, they are not available for many regions of the world. Thus a major benefit of statistical downscaling is that it can easily be applied to multi-model ensembles or to the latest climate scenarios for any location that has a few decades of adequate climate data. In other words, it does a better job of keeping up to date and capturing variability among model frameworks. The Intergovernmental Panel on Climate Change (IPCC 2007a) concluded that RCMs and SD are comparable in terms of their ability to simulate reality. Unfortunately, statistical and dynamical methods sometimes give significantly different projections for the future, highlighting the uncertainty about how a model's ability to simulate the past or present relates to its ability to project the

**TABLE 7.1** Summary of some strengths and weaknesses of three common categories of climate model. For a more detailed and technical comparison, see Randall et al. 2007.

| *Model type* | *Advantages* | *Disadvantages* |
|---|---|---|
| General Circulation Models (GCM or AOGCM) | Necessary data are easily available | Provide information only at large spatial scales |
| | Can address large-scale responses to anthropogenic forcing | Require relatively high computational time and power |
| | Projections from multiple sources are easily and freely available | |
| Regional models | More skilled at projecting weather extremes than GCMs | Require relatively high computational time and power |
| | Can provide information at a fine spatial and temporal scale | Dependent on input from GCMs, so affected by any bias/inaccuracy |
| Statistical downscaling | Requires relatively little computational time and power | Assumes relationships among large- and small-scale climate variables will not change |
| | Can provide information at a fine spatial and temporal scale | Requires observational climate data spanning the range of variability |
| | Easy to apply to multiple GCMs and to update as GCMs are updated | Dependent on input from GCMs, so affected by any bias/inaccuracy |

future. A number of modelers are now working on ways to combine SD and RCM approaches.

### *Judging the Reliability of Models or Projections*

The performance or reliability of a climate model is evaluated in a number of ways. Traditionally, models are tested by investigating their ability to simulate climate patterns covering the time period for which we have instrumental records, or climate patterns from the paleoclimatic record. Some modelers are now testing the ability of their models to make daily or yearly forecasts. Models are generally evaluated for their ability to estimate or replicate:

- Variations and trends in basic climatic variables such as temperature or precipitation;
- Short- and medium-term trends, patterns, and variability in the climate system, such as the impacts of El Niño–Southern Oscillation (ENSO) or the North Atlantic Oscillation;
- Longer-term variability such as across centuries or ice ages;
- Relationships among key variables such as temperature and humidity or cloudiness and aerosols; and
- The observed effects of a range of forcings, such as the injection of massive amounts of aerosols into the atmosphere by a volcanic eruption.

The assumption is that the more accurately a model can simulate past or present change, the more likely it is to do a good job predicting the future. It is uncertain to what degree this assumption remains valid as we push our climate system into an unknown future. The rate and magnitude of climate change may soon exceed anything seen over the last million years or more. Because of this, modelers use a variety of approaches to investigate the range of possible future climates rather than simply relying on a single model that seems "best."

One approach is to run the same model thousands of times, slightly changing the values of parameters and internal conditions each time. This helps to address the sensitivity of model output to small changes. These are known as *perturbed physics ensembles (sensitivity analyses)*.

A second approach centers on investigating the relative contribution of random elements to climate projections. As we all have experienced, weather in the physical world has a certain degree of randomness to it. While winter in general is colder than summer, we cannot be certain whether one day will be warmer than the next, or that weather systems will behave as expected. General circulation models are designed to include this element of chaos, which means that if you run the same climate model with slightly different starting conditions, you should end up with nonidentical outputs. Rather than having smooth continuous curves for temperature, wind, and other climatic variables, model outputs, like the real world, will end up with some degree of random variability—some years will be warmer or cooler simply due to random variability. Running the same model multiple times with slightly different starting conditions helps modelers to separate the chaotic elements of weather systems from long-term climate trends. These are known as *multi-ensemble model runs.*

To get an even better sense for how realistic particular projections are, modelers may do *multi-model ensembles* that compare results from multiple runs of multiple models. If the same result appears in multiple simulations from multiple models we can feel more confident that those results reflect real features of the climate system, although there is still the possibility that we have made the same error in all the models.

For its Fourth Assessment Report, the IPCC coordinated an ambitious multi-model ensemble known as Climate Model Intercomparison Project 3 that compared close to thirty general circulation models from multiple modeling centers. This has provided a stronger sense of which climate projections are robust and which are less so, based on model performance. Highly robust projections include increasing surface temperatures as a function of increasing greenhouse gas concentrations (whether from human or natural causes) and changes in water vapor with temperature. Less robust projections include changes in ENSO or hurricanes as a result of climate forcings. That is, different models give very different, sometimes opposing projections for how hurricanes or ENSO will change, indicating that we do not yet fully understand the physics of these phenomena or we are not modeling all processes, factors, and interactions that affect them.

The degree of certainty for model outputs varies spatially and for different elements of the climate system. At the scale of whole continents, for instance, there is considerable confidence that AOGCMs can provide good quantitative predictions for future temperatures (Randall et al. 2007). At smaller spatial scales, the errors in AOGCM outputs become larger. These models also do a generally poor job at simulating climate features such as tropical precipitation, extreme weather events, and the timing of shorter-term climate phenomena such as ENSO, and are less adept at dealing with the effects of varied topography and land use. These problems arise from difficulties in explicitly representing some important small-scale phenomena in models, and the high degree of uncertainty around clouds.

## Biological Models

Understanding what climate change may mean for Earth's ecosystems clearly requires looking not just at how the climate may change, but at how those changes may affect biological or ecological systems. Building on observational and experimental data, scientists have developed a broad range of mathematical and statistical approaches to modeling ecosystem or community dynamics. Models may focus on biological phenomena ranging from individual behavior to the flow of material, energy, or anything else through entire systems. Like climate models, biological models may take an empirical approach (using statistical relationships among variables of interest to project the future behavior of the system), a mathematical approach (describing biological

**BOX 7.2 USING AGENT-BASED MODELS IN ASSESSING VULNERABILITY**

A climate change vulnerability assessment in the Philippines illustrates the value of combining a range of modeling approaches (Acosta-Michlik 2004). First, a set of indicators was used to map climate change vulnerability of environmental and social systems across the entire country at the province level. This broad assessment identified Tanauan City as highly vulnerable. A second assessment used profile-based modeling to investigate the vulnerability of a subset of the forty-eight *barangays* (roughly equivalent to villages or wards) within Tanauan City. A cluster analysis of the social and economic attributes of farmers within each *barangay*, as well as their views on globalization and global change, revealed four distinct groups with distinct vulnerabilities: traditional farmers, subsistence farmers, diversified farmers, and commercial farmers. Researchers then used data from these assessments to create an agent-based model combining socio-economic and biophysical attributes of agents' environments, globally driven economic and climatic changes in those environments, and the behavior of agents (in this case, farmers) in response to their environment and changes in it. Vulnerability in this model incorporates cognitive processes (thought and reasoning) as well as exposure, sensitivity, and adaptive capacity. Commercial and diversified farmers were best able to adapt, and often acted alone. Traditional and subsistence farmers acted only after interacting with others, and how rapidly and effectively they adapted depended on the quality of their social network. The more connected a farmer was to adaptive farmers, the more rapidly he or she adapted. Using an agent-based model revealed that while engineered adaptation actions such as irrigation infrastructure and hardier crop varieties were important for reducing vulnerability, strong social networks such as farmer cooperatives were equally essential for reducing the vulnerability of traditional and subsistence farmers.

processes using mathematical equations such as differential equations), or a combination of the two.

### *Agent-Based Modeling*

Agent-based models (ABMs), also known as individual-based models, approach a system as a set of independent agents that make decisions about what to do in response to their environmental or social context. An agent may be a human, animal, cell, company, or anything that is capable of autonomous actions. Multiple types of agents may be represented within any given model, and sets of agents may be nested within other agents, such as individuals within populations within communities. Agent-based models have been used to explore a wide range of issues, including coral reef restoration (Sleeman et al. 2005), the response of fish populations to climate change (Charles et al. 2008), and human community responses to changing climate (Berman et al. 2004).

Agent-based models have several benefits when it comes to climate change adaptation. One is their ability to capture emergent properties, that is, system-wide characteristics that develop as a result of interactions among individuals within the system. They do not assume that a system exists in a stable state or has any inherent equilibrium, and explicitly allow for changes in space and time as agents respond to and change their environment. This makes it possible to explore how individual behaviors and decisions may lead to unexpected changes in the system as a whole, and to identify lever points where intervention or change would have more powerful effects. An important caveat to consider before leaping into ABMs is that their usefulness depends on accurately knowing the rules by which agents operate. If such rules cannot be determined, or if it is likely that external forces will change the rules governing agent behavior, the utility of ABMs decreases.

Because systems of agents are complex, ABMs are best seen as a way to explore the influence of individual behaviors and relationships on the system as a whole rather than as a way to make firm predictions about the future. For instance, they may help to explain how agents fail to follow conventional preference theory and end up making "irrational" choices as a result of social norms and information flow, or how a seemingly inferior technology may become more widely used than a superior one. They can be particularly useful in understanding new or emerging systems in which there are no good equations for explaining the behavior of the system as a whole, but there is some information on the behavior of agents within the system. This can be important in addressing adaptation, since in many regions climate change is pushing human and ecological systems into unknown territory.

An ABM establishes a set of decision-making rules and a set of behaviors for each agent it contains based on field studies or expert knowledge. An ABM of farmer use of seasonal climate forecasts in southern Africa, for instance, used results from sociological work with communities to determine that farmer agents in the model benefited from using climate forecasts if and only if those forecasts had been correct for at least three years. In the simplest models, each agent within a model is confronted with a set of data (i.e., it perceives the environment in which it exists), responds using the established decision rules, and generates new data (i.e., takes action). In generating new data (taking action), agents change the environment, both their own and that of other agents within the system. Thus environmental change can result from the action of agents as well as from external forces such as climate change. More sophisticated ABMs allow agents to evolve, that is, to create new behaviors that were not part of the original package.

As another example, consider an ABM of the response of a small Arctic community to changes in climate, tourism markets, and government spending on public services (Berman et al. 2004). The agents, in this case individuals nested within households, had a range of attributes such as education, wage employment, gender, and subsistence consumption targets. Based on these attributes and a set of decision rules, agents would decide whether to engage in activities related to hunting, sharing, moving, or working for pay. For instance, each household is periodically given the opportunity to participate in a hunt. Their decision is influenced by household resources, unmet subsistence need, how many caribou have been seen, and how accessible the hunting areas are. This model was used to explore a range of futures including various combinations of climate change, mass tourism, ecotourism, and government services. It turns out that decentralized risk-sharing behaviors such as communal hunting or sharing harvests made communities resilient to a range of economic changes, but were insufficient to compensate for higher levels of climate-related changes in caribou distribution and population size.

Agent-based studies can also be useful to predict or explain the failure of adaptive measures. Lansing (1991, as cited in Patt and Siebenhüner 2005) used an ABM to explain why a government-mandated switch to a uniform planting schedule with a high-yield rice variety in Bali ended up producing lower yields than the traditional temple-controlled planting and irrigation schemes the program replaced. An ABM (or at least a more nuanced policy) might have prevented the massive bankruptcy in northern Peru resulting from a government program encouraging farmers to shift from farming cotton to rice in anticipation of a strong El Niño in the 1990s. Although conditions that year were indeed more favorable for rice than cotton, so many farmers shifted to rice cultivation that overproduction of rice led to a massive drop in price that bankrupted many farmers (Remy 1998, as cited in Trigoso 2007).

### BOX 7.3 TEMPORAL GIS

Geographic Information Systems were designed primarily to address spatial questions, but are increasingly being used to address temporal questions as well. Approaches to time in GIS generally involve creating a snapshot or timeslice by time-stamping individual geographic layers (e.g., maps of a species' distribution through time) or spatial objects (e.g., particular place-based events or processes). Temporal GIS (TGIS) can also be combined with agent-based models to allow different response rules at different levels of geographic dynamics (e.g., individual organisms versus herds). Temporal GIS is not particularly good if the relationship among layers or objects is itself the focus of study, but when these relationships are well understood TGIS can be useful for addressing questions such as what types of changes have happened, where and when changes took place, and what changes will happen in the future, given a particular scenario.

In addition to its modeling capabilities, GIS can be an excellent tool for stakeholder engagement. It can facilitate the creation of vivid visual representations of change through time that can illustrate climate change concepts and consequences, and support community mapping projects that allow stakeholders to develop graphical representations of their sense of place and values.

### *Species-Based Models*

Much of past conservation and management has been based on species distributions. Although we do not know exactly how climate change will affect species distributions, we know that it already has (e.g., Parmesan 2006) and will continue to do so. Understanding how species ranges will shift over time is clearly essential for species-based conservation or management plans and for understanding how communities, ecosystems, and ecosystem services will change. Predicting species response to climate change allows us to assess adaptation potential and extinction risk and develop appropriate conservation strategies.

The most commonly employed approach to exploring the effect of climate change on species distribution is *climate envelope modeling*, also known as bioclimatic or niche-based modeling. Such models assume that current species distribution is determined primarily by climatic factors, and that future distribution can therefore be determined by projecting changes in a species' "climate envelope" over time. Models vary in terms of how they determine a species' climate envelope as well as in the assumptions they make about dispersal, ranging from no dispersal to unlimited dispersal. These differences can lead to strikingly different projections for the same species even when the starting data are identical. For instance, a comparison of nine common climate envelope models using four plant species found that projected changes in suitable climate space differed in both magnitude and direction for three of the four species, typically ranging between a 200 to 300 percent gain and a 90 to 100 percent loss in species range size (Pearson et al. 2006).

Climate envelope models have been the subject of much debate, since climatic factors are clearly not the only and in many cases not the most important factors determining species ranges. Interactions with other species, the availability of appropriate soil types, and human activities can all influence where a species occurs. Indeed, one study found that climate-species associations were no better than chance at describing the distribution of 68 of 100 bird species included in the study (Beale et al. 2008). Nonetheless, the relative simplicity of the models and of the data they require have made them popular, and they can be a useful starting point for exploring plausible future species distributions.

### *Ecosystem Models*

Ecosystem models incorporate interactions among multiple species and their environment, and can be used for exploring not just shifts in species abundance but also changes in ecosystem productivity and biomass. Dynamic global vegetation modeling, for instance, combines biological, geological, and chemical information with topographic and climatic data, and has been used at a variety of scales to project vegetation changes under future climate scenarios. In many cases, incorporating nonclimatic variables leads to significantly smaller projected future range sizes relative to climate-only projections (e.g., Preston et al. 2008).

Two basic categories of systems models have typically dominated the marine field: biogeochemical and fish-centered. Historically, the most common biogeochemical models, known as nitrogen-phytoplankton-zooplankton (NPZ) models, focused on the flux of carbon between phytoplankton and zooplankton, sometimes expanding to include detritus, dissolved nutrients, and bacteria as well. Such models have been coupled with hydrodynamic models to account for factors such as temperature, currents, or salinity. While NPZ models and their variants focus on lower trophic levels, fish-centered models focus on higher trophic levels. More complex fish-centered models, such as multispecies virtual population analyses, incorporate equations for survival, catch, predatory interactions, and changes in interactions with growth. These models typically require quantities and types of data that are rarely available for species that are not commercially harvested.

There have been a range of efforts to couple lower and higher trophic level models, with modeled links based on processes such as predation, spawning, and excretion. Most coupled models do not have truly bidirectional interactions, that is, any given process runs only from high to low or low to high trophic levels but not both. Examples of multilevel models include box-structured models such as Ecopath with Ecosim that are built around the flow between compartments, or the Integrated Generic Bay Ecosystem model built for Port Phillip Bay, Australia, that linked a coupled physical/lower trophic level model with a model of the pelagic and benthic food web.

There has been an increasing push for so-called end-to-end models that, as illustrated in figure 7.4, include the following characteristics:

- Encompass an entire food web along with the abiotic environment in which that web exists;
- Integrate physical and biological processes across a range of scales;
- Contain true two-way interactions among their diverse components (e.g., the model captures both the effect of predators upon prey and of prey upon predators); and
- Incorporate dynamic forcing from climatic and anthropogenic effects across trophic levels.

Where climate envelope models are firmly in the "simplified but utilitarian" camp, end-to-end models, or models incorporating linkages from the top to the bottom of the food web and the environment, are on the other end of the spectrum. They link processes and organisms acting across a wide range of spatial and temporal scales, incorporate both direct and indirect effects of a range of interactions, and can easily become so complex as to be unmanageable. Some modelers try to limit complexity by exploring which processes or ecosystem components are most important for explaining observed patterns. Beginning with a model that includes *no* functional relationships among model components, they see how much the model's accuracy is improved by adding increasingly complex elements such as competition, altered reproductive success, or disease transmission as a function of population density.

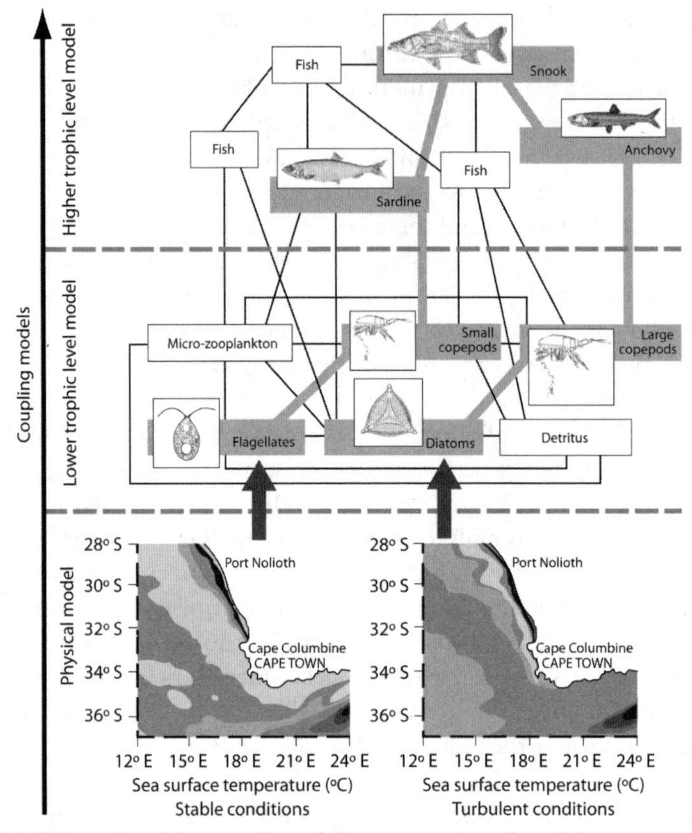


FIGURE 7.4 Schematic of the coupling of three categories of models to create an end-to-end model of a food web and its abiotic environment. The physical model (bottom) forces the lower trophic level model (middle), which is linked through two-way interactions with the higher trophic level model (top). After Cury et al. 2008.

Some scientists are calling for a new "ecosystem oceanography" that would avoid the top-down vs. bottom-up dichotomy that has typified fisheries oceanography, instead combining the two to focus on interactions at population, food web, and ecosystem levels. Thus an ecosystem oceanography model might address the combined effects of overexploitation *and* climate regime shifts.

Field and others (2006) provide an example of this combined approach for investigating climatic influences in the California Current system. Building on the Ecopath with Ecosim model, they used physical or biological time series data to drive bottom-up changes in productivity, and addressed top-down forces by altering the vulnerability of prey to predators to mimic changes in spatial distribution and production due to climate variability. They ran the model with a "constant environment" assumption, then forced it with climate indexes including those for upwelling winds, the Pacific Decadal Oscillation, and wind-derived southward transport. For some species the inclusion of interactions among species did not alter model output in any of the climate scenarios,

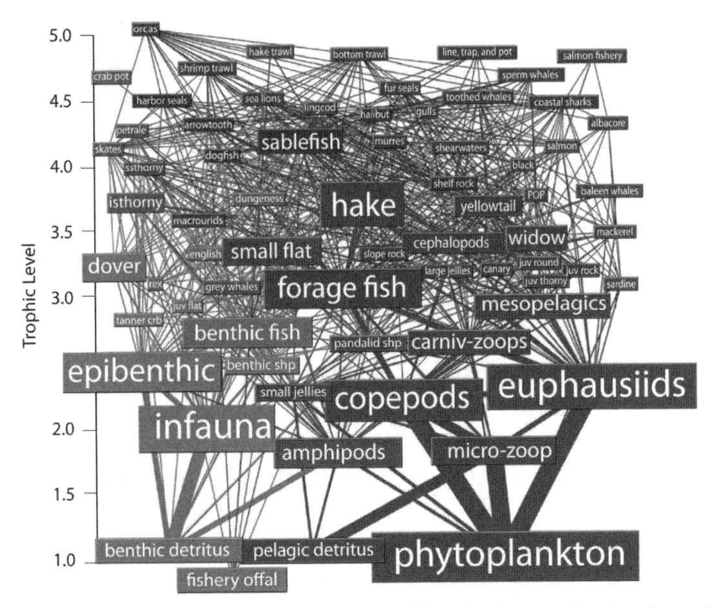


FIGURE 7.5 Visual representation of a Northern California Current food web model. The *y* axis shows estimated trophic level, while box size represents standing biomass and line width between boxes the movement of biomass from prey to predator. The drawing highlights both the daunting complexity of the system *and* the fact that it is possible to model it in a way that allows users to obtain useful information about processes as well as individual species. After Field et al. 2006.

whereas other species were significantly influenced by both top-down and bottom-up forcing. Figure 7.5 illustrates key food web constituents in this modeled system, their relative biomass, and the energy flow between them. It also illustrates how complex ecosystem models can be.

An example of ecosystem-level modeling in the terrestrial realm is an investigation of the implications of climate change for mammals in U.S. national parks (Burns et al. 2003). Based on maps of the current distribution of mammal species and ecosystem types, researchers used statistical analyses in a Geographic Information Systems (GIS) framework to quantify the association between mammal species and ecosystem type. They mapped expected ecosystem type distribution following a doubling of preindustrial carbon dioxide levels using output from the Vegetation/Ecosystem Modeling and Analysis Project (VEMAP). Models used in the VEMAP project simulated ecosystem type distribution using biogeography rules, nutrient cycling, soil characteristics, topography, fire occurrence, harvest, and climate change. Burns and her team used the equations they had developed about mammal-ecosystem associations to predict how ecosystem shifts would translate into shifts in mammalian species ranges. Based on this analysis, mammalian diversity actually increased in most parks, although many parks lost flagship species. Authors caution that their analysis did not include species interactions, or the actual ability of species to shift location over time.

While the incorporation of a wide range of ecological variables and interactions makes such ecosystem models appealing and, in theory, more realistic than simpler models, their accuracy is heavily influenced by the many assumptions that go into creating them. While ongoing testing and reevaluation of assumptions and output may help to refine any given ecosystem model, their results should, as with all models, be taken with a grain of salt.

## Sector-Based Models

Models of climate change impacts and adaptation have been developed for a number of sectors, including agriculture, forestry, fisheries, human health, hydrology, coastal zones, economics, and socioeconomics. A comprehensive review of particular models in each of these sectors is beyond the scope of this chapter, but Dickinson (2007) provides a useful synopsis.

The benefit of sector-based models is that by limiting the focus of the model, they allow more detailed and accurate representation of the important factors for the sector in question. Sectoral models can be combined into more comprehensive cross-sectoral analyses for a particular region, or provide insights as to which factors might be most important as foci for policy or action within a given sector. The term Integrated Assessment Model (IAM) has come to refer to models that couple socioeconomic and scientific elements to assess policy options.

An example of an IAM is the Dynamic Integrated model of Climate and the Economy (DICE; Nordhaus 1992). This model allows users to assess the costs and benefits of various actions to reduce global warming, and formed the basis for the widely influential *Economics of Climate Change* (Stern 2007) assessing the global economic effects of climate change. More recently, de Bruin and others (2007) modified DICE to explicitly incorporate adaptation as well as mitigation policy choices. AD-DICE, as their model is known, allows users to compare scenarios involving no action, mitigation alone, adaptation alone, and both mitigation and adaptation action. Thus the relative costs and benefits of mitigation and adaptation can be modeled separately or together.

## Final Thoughts

Models provide a useful mechanism for exploring the relationships among various factors that influence climate change and its effects on human and natural systems. Because we cannot measure the future, models also provide an essential scientific window into what the future may bring and form the basis for climate-savvy decisions about resource management and conservation. Whether qualitative or quantitative, they can clarify our thinking about the problems at hand, focus data-gathering on critical gaps, and help to develop hypotheses that can be tested with empirical data. Yet it is critical to remember that model output is *not* the same as empirical data, and that modeled projections of the future contain significant uncertainties. The structure of the models themselves may contain errors, the data used to calibrate models may contain biases, and climate, human, and ecological systems contain enough random elements that we will never be able to predict the future with complete certainty. Given these limitations, however, model output can help to clarify the range of possible futures and thus the actions needed to address the challenges it brings.

# PART II

## *Taking Action*

Addressing climate change in your work does not mean reinventing the wheel. It *does* mean making sure you have the right wheels for the terrain. Many of the tools and approaches we currently use in conservation and resource management will remain useful in a changing world, but they may need some adjustment. In other words, we cannot just keep doing what we are already doing without taking the time to make sure that we are using these tools and approaches in ways that are really helpful in the cold light of climate change. It is a bit like building a house. You could do it with no plan and only those tools you happen to have, and you might end up with a habitable domicile. That domicile would likely have many problems, however, and you might have to tear down and rebuild sections of the house to create a functional whole. Such an approach could leave you homeless for periods of time, could result in lost items in all the chaos, and would likely cost more in the long run. A better approach would be to start with a plan, or at least a vision of how it all fits together and which tools to use where. Taking action without forethought is unlikely to achieve the desired outcome.

The preceding chapters provided a philosophy and framework for how to think about adapting conservation and resource management to climate change. The next set of chapters will explore how to put this framework into action—how to connect it with what is happening on the ground. We approach this from two angles: how to adjust existing tools and practices to maintain their effectiveness in a changing climate, and which tools and practices can help to reduce the vulnerability of species, habitats, and ecosystems to climate change.

Protected areas (chapter 8) are a commonly used conservation tool, but their continued utility is uncertain given that species and ecological communities will be shifting with the changing climate. Are there ways to make protected areas more successful in meeting their existing goals in light of climate change, or do we have to think about protected areas differently under these new conditions? In this same context, how can species management (chapter 9), including endangered species recovery plans, be made more effective as conditions and communities change? Ecological connectivity (chapter 10) can help, but not all corridors or networks confer an advantage with climate change. How can they be created and managed in a way that will make them effective? Land- and seascape degradation is widespread, making restoration (chapter 11) an active field of research and practice, in particular regarding the question of whether we should now "restore" habitats to future rather than to historic conditions. Early indications are that invasive species, pests, and diseases (chapter 12) may benefit in many climate change scenarios—outcompeting, infesting, or infecting native species stressed by changing conditions. Management of these species may require reflection on new patterns of dispersal and establishment, and rethinking definitions—should species shifting naturally in response to climate change be treated like invasives? Answering these and other adaptation questions for our own work is part science, part philosophy, and part reality.

# Chapter 8

# *Strengthening Protected Areas*

There's no place like home.
—*Dorothy in* The Wizard of Oz

Protected areas have long been viewed as a primary tool of conservation biology (Groom et al. 2006). They are conceptually simple—protect space in which species or habitats of concern can exist—and legally simple—designate space and regulate allowable uses. However, as climate change alters the conditions that allow species and habitats currently in protected areas to continue to exist there, we may need to rethink protected area utility and implementation. In particular, we need to consider, as part of our climate change adaptation strategy, ways in which protected areas are vulnerable to climate change and ways in which we can make them more robust and useful as tools of adaptation.

Protected areas are often employed to protect unique, rare combinations of species or habitat. They can also be used to simply protect "wilderness," generally taken to mean relatively pristine areas that have intrinsic value as such. In either case, the idea has been that they are fixed areas that will remain unchanged forever, or at least for the foreseeable future, except in response to natural processes such as succession or random variability in population size (Soulé 1987). This basic premise makes protected areas particularly vulnerable to climate change, as starkly illustrated by the example of the National Key Deer Refuge in chapter 3. It is possible that within the next century the refuge will have no suitable habitat for its namesake species thanks to sea-level rise (fig. 8.1). This calls into question the future of the National Key Deer Refuge as a

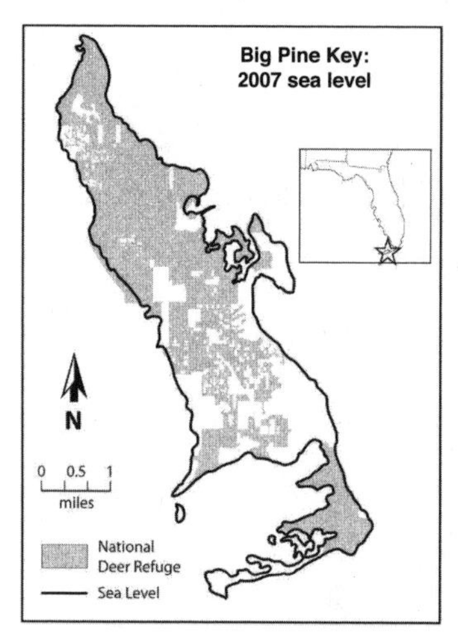


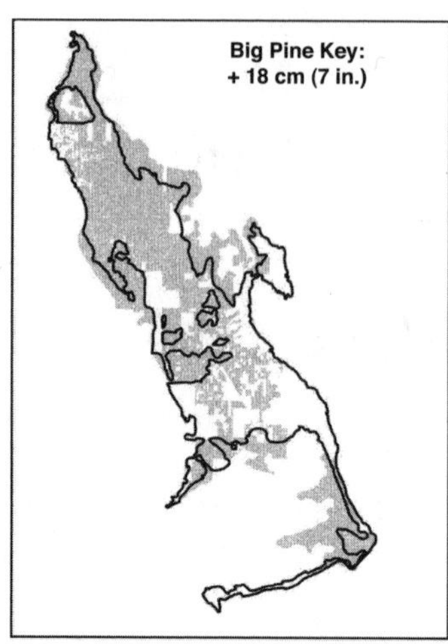


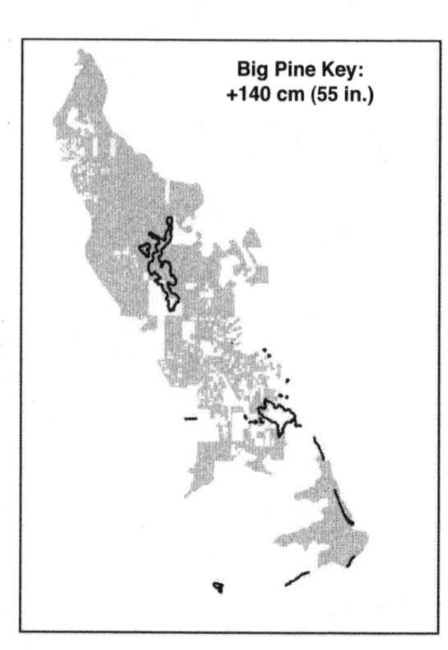


FIGURE 8.1 Habitat loss for Big Pine Key's National Deer Refuge as a function of sea-level rise for best- and worst-case scenarios over the next century. Based on maps provided by the Nature Conservancy.

terrestrial protected area, and its ability to meet its primary goal of protecting the remaining Key deer populations.

Given the historic goals of protected areas and the realities of climate change, what role can protected areas play in the future of natural resource conservation? Some argue that the utility of protected areas is diminished in light of climate change not only because most are spatially fixed, but also because many are regulated in ways that limit options for experimenting with adaptation options (Welch 2008). In the case of protected areas that are small or isolated by virtue of geography, land use, or other factors, species with shifting ranges can become trapped in or pushed out of the protected area. Further, human responses to climate change outside of protected areas can affect adaptation efforts within them. For example, reserve managers may plan for changes in rainfall by reintroducing beavers and restoring riparian vegetation, while water resource managers outside the reserve may plan to increase construction of dams and reservoirs that flood or dehydrate the reserve. The need for coordination among policymakers and managers is likely to increase as multiple parties try simultaneously to respond to climate change.

Still, there is a role for protected areas in climate adaptation. Existing protected areas work within the existing conservation infrastructure, which already has societal buy-in and funding. Supporters and managers of individual protected areas will likely become strong advocates for and participants in adapting to climate change and its effects as they see the places they care about being threatened. Protected areas can be managed at multiple scales, from landscapes to localities, to address a range of needs and stresses. In this chapter, we explore how the adaptation framework introduced in chapter 6 can be applied to protected areas (fig. 8.2), allowing us to reap their benefits while reducing their vulnerability to climate change.

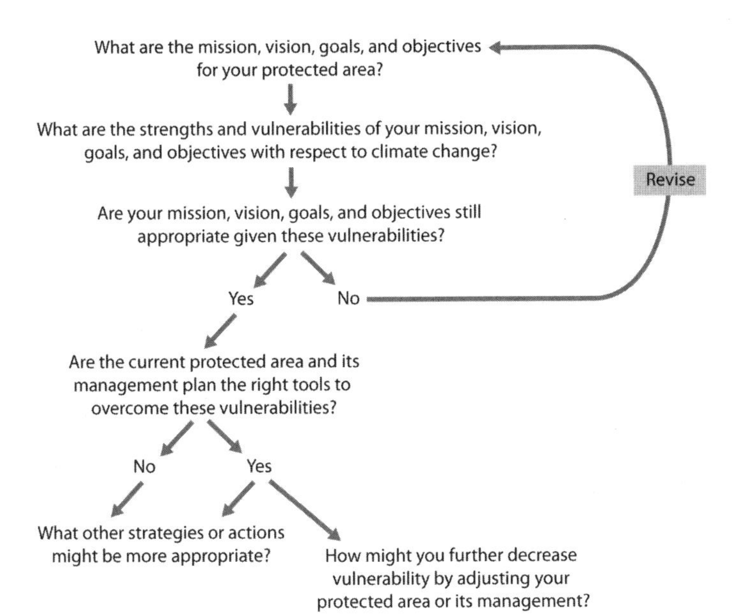


FIGURE 8.2
The application of our adaptation planning framework to protected areas.

## Goals, Objectives, and Vulnerability

Each protected area is created for a reason, and that reason is generally more than just creating a protected area. There is something—a species, habitat type, ecosystem service, or particular wilderness landscape—that creating a protected area can help to preserve, or goals it can help to meet. If the goals and objectives for your protected area are not already clearly stated, make them explicit. Begin your adaptation work by focusing on these goals and objectives.

With the goals clear, consider how they might be vulnerable to climate change. What possible changes, due to either climatic change or responses to it, could affect the feasibility of the goals? Think about both long-term trends and the variability that we will see en route, as well as the range of plausible scenarios for each. If your goal still makes sense in the face of climate change, you can move on to thinking about whether your current practices are sufficient to meet that goal in a changing world. If your goal appears highly vulnerable to climate change, take the time to consider whether it should be revised. Revising the goal or mandate of your protected area may be legally, procedurally, or emotionally difficult, but it is better than simply sticking your head in the sand and ignoring the problem.

## The Right Tools for the Job

First, ask whether your protected area is meeting its goals under existing conditions. If not, the protected area may never have been a sufficient solution, at least not by itself, and you should consider developing a different or broader strategy to meet the goal (explicitly including climate change, of course!). If the protected area *is* effectively meeting the goal, consider what is needed to maintain that effectiveness in response to current and projected changes. Will maintaining current levels of protection in the current protected area be enough? If not, how might you need to adjust the protected area, its management, or your strategy overall?

### *Protecting More*

The most often cited recommendation for adapting conservation to climate change is increasing landscape connectivity (Heller and Zavaleta 2009). Also in the top ten are other suggestions relating to protected area design such as increasing the number or size of reserves, creating buffer zones around reserves, and creating networks of reserves. Despite the frequency with which these approaches are cited as adaptation strategies, there has been limited evaluation of their efficacy. They are well worth considering, but you should carefully assess the size and design necessary to meet your goals before jumping in.

Expanding existing protected areas or networks can be an effective tool for anticipating or responding to species range shifts by enlarging or connecting existing protected areas or targeting new areas for protection. Planning for shifting ranges can be proactive in a general sort of way given the understanding that many species are moving poleward and up in elevation (Parmesan 2006), although not all species will respond in this way or at the same time (see chapter 10 for further discussion). Protected areas can also be created to target expected climate refugia that will become increasingly important for species whose possible ranges are shrinking as a result of climate change.

Start with available information about changing environmental conditions and biological responses to those changes, and consider using tools such as bioclimatic envelope models or vegetation modeling (discussed in chapter 7) to assess possible shifts in species ranges or biome types. Land use projection models can also help to elucidate the future availability of suitable habitat across the landscape. Combine these results with present and historic dispersal and range information to get a sense for how far or fast the species you care about may be able to move in response to climate change. This combination of models with empirical data and local knowledge can greatly strengthen vulnerability assessment and adaptation planning. Local or expert knowledge becomes particularly important when lack of data or expertise means that modeling is not even an option, yet action is still needed.

Based on the output of the scenarios (modeled or presumed), explore the need to expand existing protected areas or create new protected areas to meet your goals, keeping in mind how species will or will not be able to move among protected areas to track suitable conditions as change progresses. One solution is to support or reestablish connectivity between current and presumed future critical habitat areas through overlapping protected areas or managed connectivity (see chapter 10). The question of

**BOX 8.1 THINKING FOR THE LONG TERM**

**Confronted with the realities of climate change and its effects, some conservation practitioners have started to consider focusing on the landscape "canvas" rather than particular species. Mark Anderson of the Nature Conservancy has been trying to identify and protect places such as riparian zones, steep slopes, peninsulas, and regions with high physical heterogeneity whose physical features suggest that they are likely to support high levels of biodiversity regardless of climatic conditions. These "arenas of evolution," as Anderson calls them, depend on geology and landform, factors that will not change with climate change. The suite of species in these areas may change over time, but the belief is that these locales will always support more than an average number of species. They may thus serve as especially good sites for long-term conservation investment in our changing world.**

connectivity becomes increasingly complicated as a greater number of species are taken into consideration, or as you expand your thinking to include habitat requirements beyond just climatic factors. For example, if your focus is freshwater assemblages, you may need to establish protected areas along watersheds or catchments over climatic gradients.

Expanding the area under protection may also be important if species or processes require a greater area to function as a result of climate change. Species may need more room to forage in order to get enough food, or a larger population size to ensure adequate survival and dispersal under a variety of environmental conditions. Maintaining adequate water flow or other hydrological functions may likewise require land use restrictions over a wider area to compensate for diminishing water resources or altered precipitation patterns. Some rainforests and cloud forests are dependent on surrounding habitat condition to maintain sufficient rainfall or cloud cover. Deforestation in areas sometimes hundreds of kilometers away can reduce the amount of moisture reaching the forests, causing habitat loss even if the forest itself is adequately protected (Lawton et al. 2001). Thus buffer zones and connectivity concerns apply to climatic features as well as species.

### *Protecting Differently*

Climate change interacts with a range of environmental factors that are already challenges for conservation and resource management. It may be possible to decrease the vulnerability of a protected area to climate change by reducing or eliminating stresses likely to interact negatively with climate change (see chapters 12 through 16). These actions can be permanent, such as resetting regulatory limits for contaminants, or they can be seasonal, such as closing a coral reef to visitors during periods of elevated water temperatures that could induce coral bleaching.

Given the affection many have for local or iconic protected areas, the opportunity to engage and educate the public on climate change through discussion of protected area vulnerability should not be overlooked. Protected area managers can take action to lower their own emissions and show the public how they are doing so. The U.S. National Park Service's Climate Friendly Parks program is an excellent example of this approach. Park managers work to monitor and lower their emissions, while educating park visitors about climate change in general and the risks it presents to parks.

While the mere existence of a protected area can keep some stressors at bay, others are not so easily deterred. Many pollutants move through the air and along waterways; invasive species enter protected areas as hitchhikers on native species and humans; and climate change itself is unconcerned by a line on a map or a fence on the ground. Controlling those stresses likely to exacerbate or be exacerbated by climate change may thus require action beyond park boundaries. The mercury pollution that adversely affected birds, panthers, and other wildlife in Florida's Everglades National Park (Duvall and Barron 2000) comes from waste incineration and coal burning largely outside the park. Similarly, the pollutants that cause acid rain and lake acidification in national parks in Maine and Nova Scotia generally come from upwind sources along the Atlantic seaboard and much farther west. The silt, nutrients, and pollutants that can make coral reefs more vulnerable to bleaching, disease, and other problems come from terrestrial sources potentially hundreds or thousands of kilometers away—materials discharged into the Mississippi River in the Upper Midwest appear to be adversely affecting coral reefs off the coast of Florida. Thus park managers must engage closely with regulators, managers, and landowners beyond their parks to support the goals of the parks themselves.

While bottlenecks such as small population size or barriers to movement among habitat patches have always been important considerations for protected area efficacy, they must now be evaluated in light of climate change. Getting through bottlenecks in a changing climate will require considering shifting conditions and ranges when determining whether protected areas are too far apart; maintaining adequate population size and connectivity to support species' dispersal; and protecting species sufficiently in their existing ranges that they are around long enough to serve as a source population for range shifts (Vos et al. 2008).

### *Monitor, Monitor, Monitor*

As emphasized throughout this book, there is plenty of uncertainty in the world, including uncertainty about the trajectory of climate change, its effects, human responses to change, and the effectiveness of various adaptation options. Thus every adaptation plan must include some level of monitoring or information-gathering for it to be successful over time. Feedback from monitoring and research will be crucial for the development and refinement of effective management strategies. Possibilities include targeted experiments or observations to refine approaches and models, or monitoring for early warning indicators that alert managers if adaptation efforts are not working (Coenen et al. 2008).

## Innovation in Protected Area Design

The past few years have seen some new thinking about how to make protected areas more effective in response to climate change, including a few ideas that change the way we think about protected areas—movable protected areas, very large protected areas, and protecting the canvas to let nature take its course.

A number of practitioners have begun to reexamine the premise that protected areas must be in a fixed location forever. Tools like buffers, corridors, mixed-use zones, and creating a more permeable landscape can make protected areas less isolated, but it may be that in the face of rapid changes the most effective approach would be to make protected areas themselves less static. Movable protected areas would allow such areas to track the conditions that support the species or features they were created to preserve, for instance responding to the shifting range of a target species or the relocation of a physical process (e.g., coastal upwelling). Although the idea of making protected areas mobile may seem far-fetched and fraught with obstacles (e.g., the need to appease private landowners, the slow process of protected area creation, lack of certainty about effectiveness), this is essentially what regularly updating critical habitat designations for

### BOX 8.2 MOVABLE NETWORKS OF PROTECTED AREAS: WAVE OF THE FUTURE?

Dr. Dee Boersma of the University of Washington has been studying the Magellanic penguins of Punta Tombo, Argentina, for decades and watching their populations decline due to a host of stresses. While the population arrived in the region less than a century ago, it may not be there much longer thanks to climate-related changes in food availability. In the 1960s an estimated 300,000 to 400,000 penguins lived in this colony; today there are only 200,000 (Boersma 2008). These penguins are traveling progressively farther to feed as ocean productivity and their prey decline due to a combination of climate change and increasing fishing pressure. Unfortunately, when Magellanic penguin parents travel farther to forage, their reproductive success diminishes. If a penguin can find sufficient food within 70 kilometers of its nest, it can usually fledge two chicks. Expand that to between 70 and 180 kilometers, and success drops to just one chick. As the distance increases, it is more and more likely that no chicks will fledge. Today penguins travel 60 kilometers farther than they did a decade ago. Penguins also seem to be colonizing areas outside of their historic breeding range, presumably to reduce the distance they must travel to get to feeding areas at sea. The largest colony for this species, Punta Tombo, is protected by a provincial reserve, but as penguins move their colonies out onto private lands they lose that protection. One conservation option being discussed is the creation of movable reserves that track the penguins in their wintering as well as their breeding area. Fishing zoning to protect penguin feeding grounds around Punta Tombo is already in place but will need to be expanded to benefit new colonies as they move up the coast.

endangered species ideally accomplishes. The U.S. Marine Mammal Protection Act mandates a no-go area around marine mammals wherever they may be. By integrating such ideas more broadly into experimental adaptive management efforts we might be able to test, learn, and adjust to get the new mix right as climate change regularly changes the playing field for a protected area (Jessen and Patton 2008).

Another concept to consider is protecting ecological and environmental processes rather than just particular places. For example, to successfully protect wetlands we must also protect the hydrology that gets water to them. To protect rich coastal fisheries we may need to protect the processes that bring just the right level of nutrients into the system. This philosophy can be expanded even more broadly to consider protecting a "canvas" rather than the parts. In this paradigm the goal is to create an area where nature has the freedom to adapt to climate change in its own way with minimal human interference. Rather than protecting for targeted objectives, the goal becomes to support whatever should come to pass in that area. The challenge with this approach is deciding just how hands-off to be. Should we sit by while invasive species take over our designated no-interference zone? While warming temperatures lead to massive disease or pest outbreaks?

Very large protected areas have often been thought of in terms of networks of protected areas or matrix management (planning that includes the areas outside of the protected area) across a landscape or seascape. However, in response to climate change there have been even more innovative suggestions. In 2009, Helen Phillips of Natural

### BOX 8.3 AMERICAN MARTENS IN QUEBEC

Recent efforts to create a reserve around American marten breeding habitat in Quebec's boreal forest incorporated a host of disturbances and interactions, including climate change, logging, and fires. Modelers compared the projected efficacy of static protected areas during the next 200 years to that of protected areas whose location was shifted every fifty years over that same period (Rayfield et al. 2008). They also modeled a hybrid approach, a core protected area with a dynamic peripheral protected area. While dynamic protected areas supported more adequate home ranges and a higher density of habitat, they did not increase the overall amount of available habitat as compared to the static protected areas. In fact, the biggest limitation for the dynamic model was the dearth of suitable relocation options: the habitat just wasn't there.

This exercise suggests some factors to keep in mind for developing dynamic protected area systems. First, determine what criteria will be used to instigate moving the protected area, for example identifying a threshold of suitable habitat loss from existing protected areas. Second, target areas of substantial size outside existing protected areas that can become suitable habitat in the medium term. This could include cultivating forests so they become sufficiently mature. Third, remember that in at least some cases, small dynamic protected areas cannot take the place of large protected areas.

**TABLE 8.1** Four types of protected areas and ideas for making them more robust

*Traditional protected areas:* Discretely delineated areas protected by not allowing certain activities within their boundaries.

- Make them larger
- Make them more heterogeneous or diverse; include novel or dynamic assemblages at multiple scales
- Rethink what is considered exotic, invasive, or at-risk
- Include climate refugia
- Ensure viability through redundancy
- Plan for a range of future conditions in the protected area
- Plan for both chronic (long-term climate change) and acute (extreme events) conditions
- Include monitoring to assess how change is occurring (climatologically and biologically)

*Conservation matrices/networks:* Working with protected areas as part of the larger land- or seascape. Can foster semi-autonomous adaptation across this land/seascape.

- Encompass heterogeneous habitats to anticipate new climate conditions
- Include overlap of old and new climate conditions across the network
- Employ corridors and mixed-use space for connectivity
- Maintain dispersal potential by reducing stresses (barriers, pollution, and so on) in matrix area
- Protect for resilience rather than components

*"Flexible" protected areas:* Protected areas that shift over time or occur episodically (seasonal or event based). This approach can be used to protect resources for hunting or fisheries management, respond to coral bleaching, and shift breeding habitat (see box 8.3).

- Use fuller protection as areas become important spatially or temporally
- Rotate closures around existing management units to provide additional protection
- Create easements (temporary rights to land or resource use) that can be closed during times of stress or shifting needs

*Targeted refugia:* Protection of locations that are more climatically stable

- Identify climatically stable areas containing species of interest that could serve as source populations for other protected areas
- Include areas within the protected area network whose climate overlaps with future climate conditions; these allow species to make climatic transitions more slowly

England suggested managing all of the United Kingdom as a national park (Gray 2009). This would preclude creating new national parks or reserves; the focus would be on creating more wildlife-friendly habitat between reserves. This might include requiring farmers to leave more land for wild birds and mammals, and enhancing greenbelts. While this sounds very similar to matrix management, it is taking it to a new scale in which an entire country plans around a common vision of conservation management, rather than protecting some areas and paying little or no attention to others.

## The Time Is Now

Timing matters when it comes to planning and managing protected areas for climate change. Generally speaking, taking action sooner will be more successful than waiting until later, since changes currently taking place due to human activities and climate change may preclude options for successful species and ecosystem response to climate change.

Opting to gather more information about climate change while continuing to manage and protect nature in the same old way may seem like a cautious strategy because no dramatic new action is being taken, but it could in fact be quite high risk. It is a bit like trying to measure the exact speed, size, and trajectory of a train that is hurtling directly toward you rather than thinking about how to get out of the way. Taking some early action while employing flexible strategies such as robust decision making or scenario planning (discussed further in chapters 6 and 16) may leave us with a better suite of options in the future.

## A Habitat Perspective

Although the ideas presented thus far are generally applicable, they play out somewhat differently in different biomes. Below we briefly address a few idiosyncratic elements of spatial protection in a range of situations.

### *Marine*

Differences between land and sea make marine conservation and management quite different from their landlocked equivalents, and likewise pose unique challenges when it comes to adapting to climate change. Increasing atmospheric carbon dioxide is changing not just water temperature but also salinity, pH, currents, and any number of aspects of the medium in which marine creatures spend their lives. Marine species are shifting ranges more rapidly than terrestrial species (Sorte et al. 2010), as well as responding to changes in pH, which can inhibit fertilization, and shell formation and, for many highly active swimmers like the Humboldt squid, limit oxygen uptake. Oxygen levels in the water itself are changing, both in the form of large low-oxygen "dead zones" on the ocean floor and in the form of an expanding midwater low-oxygen zone.

Because of this variety of truly critical environmental variables, and because it is more difficult to map out conditions in the ocean than on land, identifying refugia for targeted protection may be trickier.

Marine protected areas (MPAs) operate on basically the same principle as other protected areas. As species abundance and distribution change, marine managers will need to create responsive programs for continuing to meet conservation or management goals. This may include flexible MPA boundaries or the creation of new MPAs, as well as large-scale management approaches that include the areas outside of MPAs.

An excellent example of this sort of large-scale approach comes from Australia's Great Barrier Reef Marine Park Authority (GBRMPA). Having determined in the mid-1990s that established zoning was not likely to afford enough protection for the reef, GBRMPA decided to take a comprehensive approach to zoning throughout the area. Well aware of the diversity of uses supported by the reef and the strong feelings of many Australians about it, GBRMPA involved stakeholders at every level of the process and from beginning to end. The result was eight different zoning categories ranging from no-go areas where access is allowed only by special permit to areas where a wide range of activities including commercial trawling, traditional harvest, and tourism are allowed (fig. 8.3). While the rezoning was not focused on climate change adaptation, it

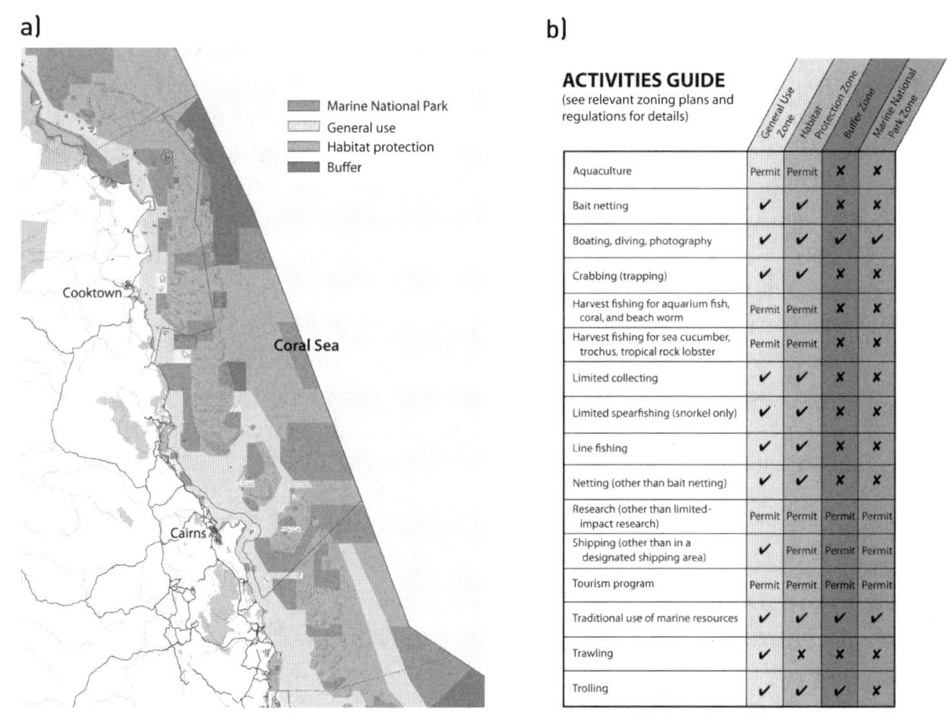


**ACTIVITIES GUIDE**
(see relevant zoning plans and regulations for details)

| | General Use Zone | Habitat Protection Zone | Buffer Zone | Marine National Park Zone |
|---|---|---|---|---|
| Aquaculture | Permit | Permit | ✘ | ✘ |
| Bait netting | ✔ | ✔ | ✘ | ✘ |
| Boating, diving, photography | ✔ | ✔ | ✔ | ✔ |
| Crabbing (trapping) | ✔ | ✔ | ✘ | ✘ |
| Harvest fishing for aquarium fish, coral, and beach worm | Permit | Permit | ✘ | ✘ |
| Harvest fishing for sea cucumber, trochus, tropical rock lobster | Permit | Permit | ✘ | ✘ |
| Limited collecting | ✔ | ✔ | ✘ | ✘ |
| Limited spearfishing (snorkel only) | ✔ | ✔ | ✘ | ✘ |
| Line fishing | ✔ | ✔ | ✘ | ✘ |
| Netting (other than bait netting) | ✔ | ✔ | ✘ | ✘ |
| Research (other than limited-impact research) | Permit | Permit | Permit | Permit |
| Shipping (other than in a designated shipping area) | ✔ | Permit | Permit | Permit |
| Tourism program | Permit | Permit | Permit | Permit |
| Traditional use of marine resources | ✔ | ✔ | ✔ | ✔ |
| Trawling | ✔ | ✘ | ✘ | ✘ |
| Trolling | ✔ | ✔ | ✔ | ✘ |

FIGURE 8.3 Spatial planning along Australia's Great Barrier Reef. This simplified map shows allowable uses for one section of the Great Barrier Reef (a), and the allowable activities in those zones (b). GBRMPA identified sixteen different uses of the area around the reef, and created eight zones (four of which are shown here) where different combinations of activity are allowed.

turned out to be quite useful toward that end. Some of the aspects likely to support resilience in the face of climate change are having a network of strictly protected areas within a broader matrix of varying protection, maintaining ecological functions, and creating ecological safety margins to protect against disaster.

### *Freshwater*

Freshwater areas are heavily used the world over and tend to have limited representation or protection in protected areas. Because of human reliance on freshwater for a range of uses and its limited availability, competition for this resource is often more fierce than for land. Humans will certainly act to maintain water supplies in the face of climate change, and proposed actions such as increased irrigation or more water diversion or retention projects (aqueducts and dams) will conflict with actions proposed to decrease the vulnerability of freshwater systems themselves. For instance, removing dams to allow species to move upriver to potentially cooler microhabitats is a frequent suggestion for adapting fish conservation to climate change, but conflicts with calls for increased numbers of reservoirs to ensure adequate summer water supplies in drought-prone areas. Likewise, ensuring adequate water flow within and outside protected areas will become increasingly difficult as overall supplies drop. Managers whose goals include maintaining healthy freshwater habitat within their protected area will need to work even harder to ensure that water allocation outside the reserves does not make it impossible to meet those goals.

Other elements of creating climate-savvy freshwater protected areas may be somewhat less contentious, for instance prioritizing cold-water refugia for protection, protecting and restoring riparian corridors to help keep rivers and streams cool, being increasingly vigilant with regard to invasive species, or protecting recharge zones (Pittock et al. 2008). There are even some win-win propositions, such as restoring floodplains or wetlands to reduce flood risk.

Riparian corridors are particularly interesting in terms of adapting to climate change as they essentially create linear protected areas that span altitudinal and therefore climatic gradients. In addition to benefits for aquatic species, a system of protected riparian corridors might also provide an avenue for some terrestrial species to track appropriate climate conditions.

### *Terrestrial*

Terrestrial ecosystems have the benefit of being more familiar and intuitive to people than marine or freshwater systems, but also the challenge of being where most human activity takes place. Habitat fragmentation, pollution, and diminished population sizes can limit the ability of species to respond to current climatic changes as they have to past ones, for instance through migration, range shifts, or evolutionary adaptation. The presence of extensive infrastructure makes more radical ideas such as moveable protected areas potentially difficult to implement.

On the other hand, our ability to see and measure climate change and its effects across terrestrial landscapes makes it potentially easier to identify and target climate gradients and refugia. It is often easier for conservation and resource planners to develop management strategies for terrestrial systems, with or without consideration of climate change. This may make development of holistic approaches to climate-aware management easier from a terrestrial perspective, including integration across sectors (human and natural systems) and habitat types (terrestrial with freshwater and marine interfaces).

### *Small Islands*

Small islands, and even small island chains, are typically highly vulnerable to climate change. While islands range from low-lying atolls to steep mountains and volcanoes, the amount of available habitat is small, making ongoing changes like sea-level rise and increasing temperatures or stochastic singular events like intense storms equally challenging for the long-term success of established systems. For isolated islands, there is simply nowhere to go and nowhere to come back from for anything other than species capable of traveling long distances by air or water. Protected areas on small islands in some cases already encompass the entirety of an island or exist in the highest-elevation regions, so there is limited potential for redesign, expansion, or connectivity (Gerlach 2008). Sea-level rise could submerge them entirely, or inundate or otherwise damage coastal wetlands and other coastal habitat types. Small islands often contain particularly climate-vulnerable habitats, like mist forests, which will likely be lost with climate change. Very few options for improving protected areas on small islands have been suggested.

### *High-Latitude Regions*

Some of the most rapid climate change is occurring in high-latitude regions of the world, giving managers there much less time to figure out what to do. Most of the Arctic and the Antarctic Peninsula have experienced temperature increases at least twice as great as the global average, as well as substantial changes in the amount, type, and timing of precipitation (Anisimov et al. 2007). Melting permafrost is rapidly changing the nature of high-latitude regions, and loss of sea ice threatens the very survival of a number of species that are absolutely dependent on sea ice for survival. In the Northern Hemisphere, species ranges are shifting rapidly, posing the twin challenges of incorporating new species into management or conservation plans and combating increasing numbers of nuisance species (pests, diseases, and invasives).

With ice being the primary feature of high-latitude environments historically, climate-savvy protected area design likely needs to include protected habitat that will remain frozen longer (refugia), connectivity between these refugia, and space for the system to change what it relies on as even those refugia disappear. This probably means very large protected areas. Some conversation has emerged of a whole Arctic protected area along the lines of the Antarctic Treaty protections. In Antarctica there may be a need to reduce access or increase management of tourism in protected areas to increase resilience of the region to climate change.

## Final Thoughts

Despite the potential challenges posed by climate change, we should not abandon existing protected areas or start relocating them wholesale to the locations we predict will be useful in the future. If the past is any guide, at least some species will stay put during periods of climate change and will benefit from the continued presence of existing protected areas. Also, establishing new protected areas takes a long time politically and ecologically, and there is no guarantee of success on either front. In many regions, there are few natural areas near existing protected areas that remain for gazetting, making continued protection of what we have that much more important. Even if existing protected areas become climatically unable to support the assemblages for which they were originally designed, they may prove useful as habitat for new communities moving into the region or as stepping-stones as species reshuffle across the landscape.

# Chapter 9

# *Focusing on Species*

A rose is a rose is a rose.
—*Gertrude Stein*

Individual species have historically been a major focus of conservation efforts. The reasons for this are varied and include the importance of particular species for subsistence or other purposes; religious, spiritual, or ethical values; and the effects on the public of individual species being presented in isolation in zoos, movies, advertising, and educational campaigns. Numerous regulations, laws, and treaties focus on the protection or status of individual species. In the United States, the Endangered Species Act of 1973 provides "a means whereby the ecosystems upon which endangered species and threatened species depend may be conserved," where conservation is defined as "the use of all methods and procedures which are necessary to bring any endangered species or threatened species to the point at which the measures provided pursuant to this act are no longer necessary." The global Convention on International Trade in Endangered Species of Wild Fauna and Flora (CITES), which came into force in 1975, is built around the premise "that wild fauna and flora in their many beautiful and varied forms are an irreplaceable part of the natural systems of the earth which must be protected for this and the generations to come."

Climate change presents a number of challenges to species-based conservation approaches. Past ecological relationships and linkages may be broken, upending many of the assumptions used in species-based conservation. Species may no longer occur in areas where they have historically been found, and may appear in new areas. For many species, baseline data are scarce and our understanding of how climatic factors contribute to a species' distribution and well-being is often extremely limited.

## Some Ecological Considerations

Regardless of the challenges posed by climate change, the myriad reasons for focusing on species-based conservation are not likely to go away. The question then becomes how best to adapt species-based strategies to a changing world. Before getting to particular approaches, we would like to highlight a few issues worth considering in the formulation of climate-savvy species-based plans.

### *Individualistic Species Responses to Change*

During past periods of climatic change, species responded individualistically. The ranges of some expanded, while those of others contracted or shifted. Populations within some species became isolated and evolved into distinct species or subspecies, while in other cases previously separate populations reconnected and interbred. Some species have population cycles with clear links to climatic variables, while others do not (e.g. A'mar et al. 2009). Individualistic responses at the species level make predicting community or biome shifts difficult, because different community members may respond differently and entirely new community types may appear (e.g., Graham and Grimm 1990). Evolutionary, behavioral, and physiological responses to climate change as well as to unrelated environmental changes may further complicate conservation efforts. It is easier to incorporate these sorts of individualistic needs and responses when using a species- rather than community-based approach to conservation and management, although the risk is that important community-wide dynamics may be missed.

The individualistic nature of species or species groups is important in the design of conservation and management plans regardless of climate change. Mechanisms of individual movement or species dispersal have received significant attention in the connectivity literature, and studies have shown that multispecies conservation plans are less effective than single-species plans (Clark and Harvey 2002; Rahn et al. 2006).

### *Keystone Species and Ecosystem Engineers*

Some species play more important roles than others in structuring the communities or ecosystems in which they live. This idea began to crystallize in ecology following the publication of a seminal paper on how the presence or absence of predatory sea stars reshapes intertidal communities (Paine 1969), and the concept of keystone species has been expanded to include any species whose removal would cause dramatic changes in the community, be it through predation, competition, or modification of the physical or chemical environment. Species in the latter category, such as beavers or turf-forming grasses, are often termed ecosystem engineers. Recently, researchers have begun to investigate how keystone species and climatic gradients interact to determine the presence or absence of species over larger scales (fig. 9.1). Designing climate-smart conservation and management around effects on keystone species or ecosystem engineers may

**BOX 9.1 CLIMATE, PREDATION, AND SURVIVAL**

On Washington's northwest coast, California mussels survive in an intertidal band high enough to avoid predatory sea stars, but low enough that they are not scorched during low tides. On the outer coast the combination of large waves, fog, and morning low tides provides ample room for the mussels. In Puget Sound, waves are smaller and summer low tides occur near midday. Mussels cannot survive above the sea star zone and are virtually absent from the Sound. This suggests that as the climate warms throughout this mussel's range, the zone in which it exists may shrink. A similar dynamic can play out between seaweed and the limpets that graze on them on scales as small a a single boulder. On the cooler north-facingside of te boulder, seaweed can grow high enough to escape its less heat-tolerant predator; on the hotter south-facing side, there is no such refuge.

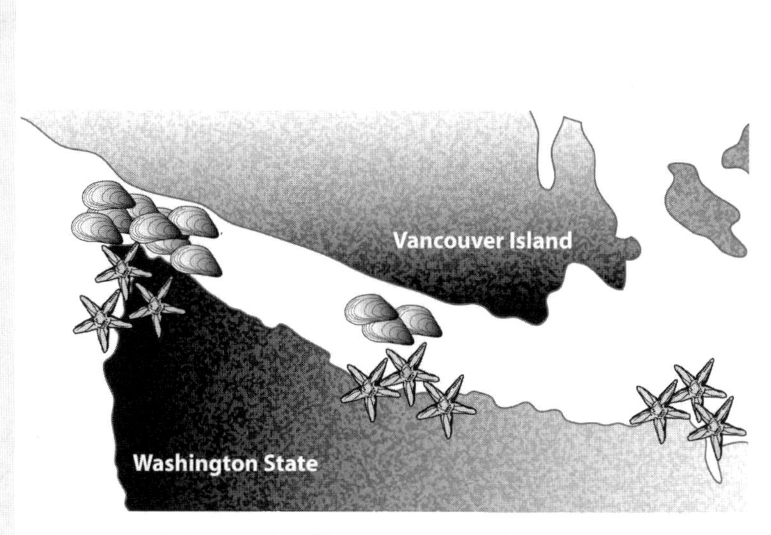


FIGURE 9.1 Schematic of how sea stars and climatic gradients combine to keep California mussels out of Puget Sound. On the outer coast (left) sea stars and mussels coexist, but increasingly warm intertidal conditions moving into the South (right) shrink te muscle's safety zone to zero (C. Harley, unpub.).

allow the use of mechanistic, species-based models and management while also serving to increase resilience at the community or ecosystem level. Further, some ecosystem engineers may actually decrease vulnerability to climate change or its effects. For instance, some salt-marsh species increase coastal accretion, potentially reducing erosion and thereby the effects of sea-level rise. By creating wetlands and micro-dams, beavers can decrease the effects of both more severe droughts and rainstorms.

### *Umbrella and Indicator Species*

Monitoring and protecting all or even a moderate proportion of species in a particular community or ecosystem can be costly and time-consuming. Two related concepts that have been suggested for limiting the number of focal species while achieving a high level of protection are umbrella species and indicator species.

An umbrella species is a species whose protection would result in protection for multiple co-occurring species as well. To be effective, an umbrella species should have range and habitat-use patterns that overlap with those of many other species. Protecting areas of old-growth forest for spotted owls, for instance, is presumed to also protect snails and salamanders within those forests. If an umbrella species is less sensitive to climate changes than species with which it co-occurs, however, protecting the umbrella species may no longer provide the broad benefits originally anticipated.

In contrast, indicator species are those whose status reflects something about the status of the habitat in which they occur. Peregrine falcons have been used as indicators of pesticide levels, and a variety of freshwater animals are regularly used as indicators of water quality in rivers, streams, and lakes. It may be possible to identify species that are particularly sensitive to various effects of climate change that may serve as early-warning indicators of problems to come. For instance, many amphibians are particularly sensitive to changes in moisture, while corals are particularly sensitive to high water temperature.

### *Migratory Species*

Focusing conservation efforts at the habitat or ecosystem level may well provide sufficient protection for some of the species that make use of the habitat, but this is not the case for migratory species or species that use different habitats at different life stages. Migratory birds may travel thousands of miles between summering and wintering grounds, using distinct habitats across their migratory pathway. Likewise, species that spend their early life stages in one habitat and their adult life in another can be protected in the wild only if all habitats used throughout their life are protected. Dragonflies, for instance, spend their larval stages in the water and their adult life on land (or in the air). While these are conservation challenges regardless of climate change, it is worth emphasizing that vulnerability assessments and adaptation plans must take into account all habitats essential to a species' existence.

## Adapting Existing Approaches to Management

Existing approaches to species-based conservation do not have to be abandoned wholesale in response to climate change. Many can be adjusted to address climate change. In some cases, simply applying existing best practice guidelines would be a step in the right direction; we focus here on efforts that would adjust or alter those practices.

### *Listing Species as Threatened or Endangered*

A number of mechanisms exist for identifying species as being particularly at risk of extinction, ranging from global (e.g., the International Union for the Conservation of Nature's Red List of Threatened Species, the Convention on the Conservation of Migratory Species of Wild Animals) to federal, state, or provincial endangered species laws or policies. Several precedents exist for considering climate change effects when making decisions about the status of species of concern. In 2001 Australia designated "loss of climatic habitat caused by anthropogenic emissions of greenhouse gases" as a threatening process under its Environment Protection and Biodiversity Conservation Act of 1999, and roughly one-third of the animals on Australia's national "endangered" list are threatened by climate change. In the United States two species of coral were listed as threatened under the Endangered Species Act in 2006 with the top three reasons for their listing—warming water, increased incidence of disease, and increasing hurricane damage—all having links to climate change. Polar bears were likewise listed as threatened in 2008 in large part due to disappearing sea ice. The IUCN recently developed a set of criteria for assessing the vulnerability of species to climate change and applied them to amphibians, birds, and warm-water reef-building corals. Between 70 percent and 80 percent of bird, coral, and amphibian species currently listed as threatened by the IUCN are susceptible to climate change, and up to 70 percent of unlisted species met the criteria for climate change susceptibility. Species that are not currently listed as threatened but have high susceptibility to climate change should be investigated further.

### *Conservation, Recovery, and Management Plans*

Most conservation, recovery, and management plans to date ignore climate change or mention it only briefly without substantial discussion of response options, although this is starting to change (Povilitis and Suckling 2010). The conservation plan for Cook Inlet beluga whales, for instance, provides a reasonable discussion of possible effects of climate change, but places it under the umbrella of natural environmental change and concludes that there are insufficient data to determine actual effects on beluga (National Marine Fisheries Service 2008). As a result there are no recommended management actions that address possible threats from climate change.

One plan that stands out for its incorporation of climate change is the Conservation Plan for the Greater Sage-grouse in Idaho (Idaho Sage-grouse Advisory Committee 2006). The plan discusses nineteen threats to greater sage-grouse in Idaho and ranks their importance at three scales: range-wide, statewide, and within designated Sage-grouse Planning Areas (SGPA). At the rangewide scale, climate change per se was not among the top ten, but three climate-sensitive factors (invasive species, wildfire, and weather) were. Statewide, climate change was listed as the ninth biggest threat. The document acknowledges that many of the threats they have listed separately are tightly interlinked, and that effects of climate change are likely to be region- and species-specific. Because of the uncertainty and variability, the committee concludes that the key to managing rangelands in the face of climate change will be to enhance resilience using a range of approaches, including reducing nonclimate stressors, restoring degraded areas with an eye toward future conditions, increasing management flexibility, and improving monitoring.

Changing conditions should also inform decisions about what qualifies as essential habitat for endangered species. This may mean designating as critical habitat areas that are not currently critical but are likely to become so in the future. The recent critical habitat ruling for the Quino checkerspot butterfly (U.S. Fish and Wildlife Service 2009; fig. 9.2) includes a thorough discussion of observed and projected changes in distribution, host plant use, and habitat fragmentation for this subspecies as a result of climate change, and incorporates this information into the final habitat designation in multiple ways. The rule cites a need to create greater connectivity between some of the core occurrence complexes and higher-elevation habitat in order to facilitate range shifts in response to climatic change, and to provide particular protection for

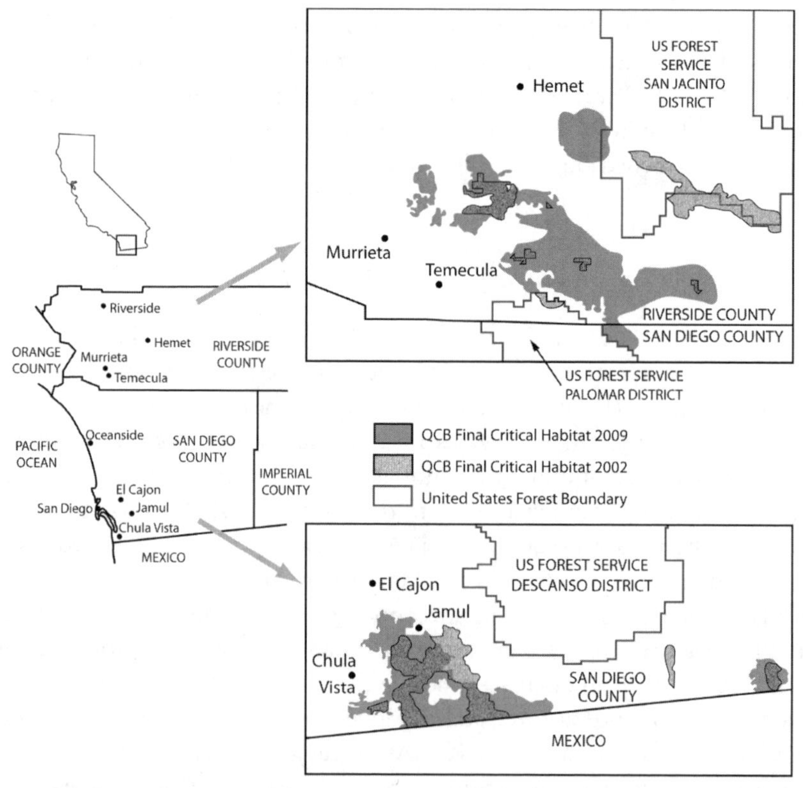


FIGURE 9.2 Map of current and formerly designated Quino checkerspot butterfly critical habit. The changes reflect the disappearance of some lower-elevation or southern populations and the inclusion of some higher-elevation or atypical environments considered important for the ability of the subspecies to adapt to climate change.

populations locally adapted to drier climates since these may have characteristics that would make them more resilient in the future.

### *Population Viability Analysis*

Population viability analysis is a well-established tool in species conservation and management, incorporating such elements as minimum viable population size, density, age structure, and so on. Although the optimal values for some elements of population viability are unlikely to be affected by climate change, others may well be sensitive to it. For instance, the minimum viable population size and density may change if population dynamics are strongly affected by climate change. If generation time decreases or reproductive success increases because of warming temperature or other changes, a smaller population size may be sufficient for viability. Conversely, if conditions become increasingly stressful, reproductive success may decline, increasing the minimum population size needed to provide a high likelihood of long-term viability. Given that evolutionary adaptation may be a critical component of viability in the face of climate change, the level of genetic variation needed to ensure viability may also be higher in a rapidly changing world.

## Supporting Nature

Natural processes such as behavioral changes, shifting ranges, and evolution can all contribute to long-term resilience of species and ecosystems in the face of climate change. Understanding and supporting these processes can provide managers and practitioners with important tools to reduce the vulnerability of species of concern. While such an approach will not always be sufficient to guarantee the survival of species in light of rapid change and habitat loss, it is a "no regrets" option that provides benefits beyond climate adaptation.

### *Range Changes*

The paleontological record is full of examples of species ranges shrinking, expanding, shifting, or disappearing in response to climatic changes, and some species ranges are already changing in response to the present period of rapid climate change. Being able to predict whether and how a species' range might shift would clearly facilitate efforts to manage that species. Bioclimatic envelope models are an effort to do just this, although as discussed in chapter 7 the results of these models should be viewed with appropriate caution. Still, such models can provide a framework within which to consider future species distributions.

Although some practitioners question the utility of static protected areas in the face of shifting species ranges, it is important to remember that not all species will undergo range shifts. For species that remain within their existing range, static protected areas may remain a viable conservation tool. For species whose ranges shift far enough to move them beyond the boundaries of existing protected areas or regulatory regions, new approaches will be needed. Connectivity strategies geared toward species of concern are one option (see chapters 9 and 10). Another is to preemptively establish regulatory frameworks in regions where species do not yet occur but are likely to in the future. For instance, if a particular species of fish currently does not occur in Canada but is likely to appear there within the next decade or two, harvest or protection regulations might be put in place before it appears to provide greater predictability for resource users and immediate protection (if warranted) when the species appears. Coordination among managers and regulatory bodies in adjacent jurisdictions could greatly enhance the success of such efforts.

A linked concept is establishing regulations that follow species wherever they go or respond rapidly to changes in species behavior or distribution. Existing examples of such approaches include the Marine Mammal Protection Act in the United States, which mandates a minimum distance between people and marine mammals, or any regulations whose requirements apply to a particular species rather than a particular place. Many fishing regulations, for example, are spatially and temporally responsive to species distribution, establishing rolling closures.

### *Evolutionary Adaptation*

Despite its foundational importance in modern biology, there is not yet a coherent framework for incorporating evolutionary processes into wildlife conservation and management. This is clearly a deficit regardless of what the climate does, but becomes particularly important when working to build climate change into conservation and management (fig. 9.3). The existence of geographic variation in temperature tolerance or other climatic sensitivity within a wide array of species suggests that genetic variation in climate-related traits exists and that evolutionary adaptation is possible. Rapid adaptive evolution in response to other anthropogenic stressors by a range of species demonstrates that in at least some cases, evolutionary responses can occur on timescales relevant to both climate change and conservation (Rice and Emery 2003). The key is figuring out when evolutionary responses can be ignored and when they are an essential consideration.

A first step toward supporting evolutionary adaptation to climate change would be to assess adaptive variation across a species' range. While populations at the warmer range edge are often dismissed as being most likely to go extinct as conditions heat up, they may be an important reservoir of genetic material for adapting to a warmer world. A less targeted approach of maintaining overall genetic diversity within and among populations may boost generalized resilience. For instance, artisanal and subsistence farmers using a greater diversity of wheat varieties typically experience more consistent yields over time, while large-scale agriculture operations dependent on a limited number of cultivars are highly vulnerable to new diseases or climatic shifts. Alternatively, managers and practitioners may take a more interventionist approach, using agricultural breeding techniques to create more resilient populations of plants and animals. In

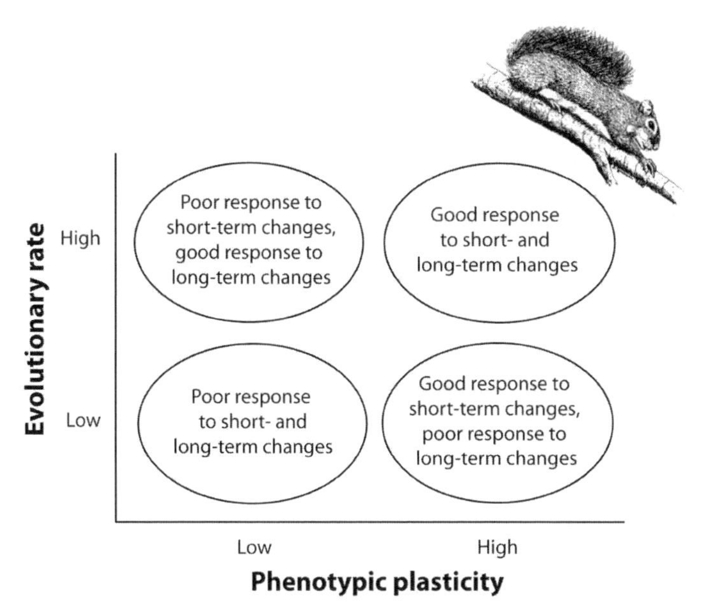


FIGURE 9.3 Phenotypic plasticity (changes that occur within an individual over its lifetime) can be effective for coping with environmental changes in the short term. Evolutionary rate, determined by factors including generation time and genetic variation within populations, does not help individuals in the short term but contributes to the ability of populations to cope with change over the longer term. The trend toward earlier pupping in red squirrels in the Yukon over the last ten years results from a combination of phenotypic plasticity and evolution. After Berteaux et al. 2004.

the southeastern United States, for instance, U.S. Geological Survey scientists worked to develop more salt-tolerant cypress that would survive increasing saltwater intrusion into coastal swamps.

The goal of evolutionarily enlightened conservation and management may be different in static versus changing situations. Local adaptation may be a desirable characteristic if conditions are relatively static, since it optimizes reproductive output and success under prevailing conditions, but if conditions are changing, strong adaptation to the status quo may become a liability rather than an asset. Regulations that limit the mixing of genes across a species' geographic range (for instance, requirements that restoration or replanting be done only with locally obtained genotypes) may need to be reassessed to determine whether they may prevent adaptive responses. It is critical to note that even in the face of climate change there are situations in which the preservation of genetic distinctions among populations may be desirable.

## Beyond Nature: Interventionist Approaches

When threats to natural resources appear severe, managers and practitioners may consider actions that go beyond simply supporting natural processes, such as deliberately

**BOX 9.2 EVIDENCE-BASED ADAPTATION**

Cattle are often viewed as a foe to wetlands. They physically disturb delicate habitats, and overgrazing can increase sedimentation. Yet cattle may actually increase the resilience of vernal pools in parts of California's Central Valley to climate change (Pyke and Marty 2005; Marty 2005): pools in areas from which cattle were excluded dried out on average fifty days earlier than those in areas with year-round grazing and had a lower diversity of native plants. By controlling the growth of plants in and around these pools, cattle decrease water loss, a factor that will likely become even more important as climate change reduces precipitation to this region. This is good news for fairy shrimp and other species that rely on this temporary resource to complete their life cycle.

moving species beyond their current range or actively directing evolution. While such strong intervention may be the only option for some species, the possibility of unanticipated undesirable consequences can be high. Managers and practitioners should make sure that they are acting based on realistic, thoughtful, and thorough cost-benefit analyses rather than simply "climageddon" panic before engaging in irreversible interventions in the natural world.

### *Assisted Migration*

Assisted migration, also known as managed relocation or assisted colonization, is the purposeful introduction of a species from where it currently occurs to an area where it currently does not. It may include introducing a species to an area where it once existed but has disappeared, to an area within its range where it was not known to occur previously, or to an area completely outside of its historical range. Because of the well-documented economic and ecological problems associated with nonnative invasive species, as well as the limited success of past reintroduction programs, many people regard assisted migration with a good deal of concern. While some have suggested that assisted migration may not be as damaging for recipient ecosystems as typical nonnative invasions because the distance over which species are moved will often be smaller, it is not clear that distance from origin correlates strongly with a species' effect on new ecosystems (Mueller and Hellmann 2008). Furthermore, even species naturally expanding their ranges have had negative effects on their new communities (Sorte et al. 2010).

The idea of managed relocation as a conservation strategy pre-dates widespread concern over climate change. The IUCN's Reintroduction Specialist Group was formed in 1988 in response to concern over habitat loss unrelated to climate change, and in 1998 the IUCN released a set of international translocation guidelines concluding that managed relocation "should be undertaken only as a last resort when no opportunities for re-introduction into the original site or range exist and only when a significant contribution to the conservation of the species will result."

The argument for managed relocation as a response to climate change is that current climate change is occurring so rapidly that many species may not be able to evolve or shift their ranges fast enough to remain in climatically suitable habitat. Furthermore, urbanization and other land use changes impede the ability of species to move across landscapes. While climate change does present a new threat to some species' persistence, the basic practical and philosophical issues related to assisted migration remain the same. When, if ever, is it justifiable to facilitate the spread of a species beyond its current range in an effort to increase its likelihood of survival? When do the needs of a single species trump the risks to the recipient community of going through with relocation? Because the risks to the recipient community are so high, managed relocation should generally be considered only as a last-ditch effort.

A number of authors have proposed risk-benefit frameworks for deciding if or when to do managed relocation (Hoegh-Guldberg et al. 2008; Richardson et al. 2009). The most basic questions are whether there is a high risk of decline or extinction, whether translocation would address any of the problems contributing to extinction risk, and whether translocation and establishment in a new location are technically feasible. If the answer to any of those questions is no, assisted migration should not be considered. If the answer to all questions is yes, a more detailed risk-benefit analysis is needed. The analysis should include effects on the populations and ecosystems from which relocated individuals are taken, effects on the ecosystems into which individuals are placed, and the political, financial, and social capital necessary to move a project forward. While some elements of the analysis rest on scientific information, others are value-driven. Davidson and Simkanin (2008) suggest that, given the high risk of disruption to recipient ecosystems, assisted migration makes sense only if the recipient ecosystem is viewed as having less conservation value than the species being relocated.

### *Captive Breeding*

Captive breeding programs are usually undertaken either as a last-ditch effort to save species whose extinction in the wild seems all but guaranteed, or as an effort by zoos or individuals to maintain their own populations of particular species. There are a handful of cases in which such programs have led to apparently successful reintroduction of a species into the wild. Przewalski's horse and the European bison were both reintroduced to the wild following intensive efforts using zoo populations. The remaining California condors and black-footed ferrets were taken from the wild when extinction seemed imminent and bred in captivity until enough individuals existed to make reintroduction to the wild possible. Success is not the norm for reintroduction efforts, however, and this fact plus the extensive resources required for such efforts makes them controversial. Climate change is likely to increase the demand for captive breeding because of increasing extinction rates in the wild, but it may also decrease the success of reintroduction efforts.

A potential unintended consequence of captive breeding and release programs is a long-term loss of evolutionary fitness (e.g., Araki et al. 2007). To the extent that captive breeding occurs under "ideal" conditions, individuals that are produced by such programs may be maladapted to the conditions they encounter in the wild. Furthermore, lack of natural selection in captive breeding programs may limit evolutionary adaptation to changing conditions if a significant proportion of a population comes from such a program. Captive breeding programs that incorporate a range of plausible climate conditions into their design (as in the salt-tolerant cypress program mentioned above) may thus experience greater long-term success.

### *Creating Artificial Refuges*

If the value of a species is particularly high, managers may wish to take more interventionist actions (for instance, assisted migration as discussed above). An intervention particular to climate change is the creation of temporary or permanent refuges from climatic stressors. For instance, shaded hatcheries have been set up near some sea turtle nesting beaches that frequently become lethally hot for the turtle eggs. In some cases, hatcheries will remain necessary as long as turtles continue to nest at that particular beach. In other cases, targeted restoration or other intervention may decrease the need for hatcheries, for instance by restoring coastal vegetation that cools nesting beach temperature. On coral reefs, people have experimented with floating cloth or sprinklers to create artificial shade for corals when bleaching risk is high.

Whether to use such interventionist approaches depends on a number of factors, including costs, feasibility, desperation, and current degree of management. In wilderness areas resistance to intervention is likely to be high, while in a landscape that is already heavily managed it will likely be lower. Costs can be minimized by establishing "self-maintaining" refuges. For instance, in the Galapagos, scientists are considering creating cooler nesting conditions for land tortoises by planting shade plants. Once established, the plants would require little attention. Likewise, restoration site design can incorporate boulders, gullies, or other permanent shade-producing features to create cooler microhabitats.

## The Question of Triage

There will never be enough resources to mount conservation campaigns for all species in need of them, with or without climate change. Nonetheless, climate change has brought renewed life to the old question of triage in conservation and management. Triage literally means sorting or selecting, and in disaster response refers to a system for prioritizing individuals for treatment. While conservation triage happens all the time in an ad hoc manner (no organization focuses equally on all species, after all), some feel that the conservation and resource management community should take a more comprehensive and conscious approach to the choices it makes.

Current regulations and policies, such as the U.S. Endangered Species Act, tend to focus on the species at greatest risk of extinction. Traditional conservation organizations have likewise focused on the most vulnerable, with the goal of preventing any species from going extinct. In the face of both limited resources and an increasing number of threats to species, there are calls to abandon "doomed" species and focus instead on species where intervention has a higher likelihood of long-term success. This approach carries multiple risks, including inappropriate abandonment of species based on insufficient data or incorrect assumptions, and should clearly be approached with great caution.

While triage criteria can be adjusted to reflect a variety of values (e.g., Lawler 2009), it is important to remember that triage systems were developed for emergency situations. They are not the only option for prioritization, and may not be appropriate as a long-term strategy (Millar et al. 2007)

## Final Thoughts

Although climate change does test traditional species-based conservation in a number of ways, it is certainly not time to abandon all hope or even all existing approaches to species-based conservation. Humans will continue to manage and protect species, so we might as well do it in the best way we can. Today, that means incorporating climate change considerations into our work. A number of existing approaches can be modified to function more effectively in a changing climate, and novel tools and approaches will doubtless arise as people respond creatively to the challenge. A pitfall to avoid is allowing desperation to lead us into actions with potentially damaging unintended consequences without taking the time to carefully assess the likelihood and severity of such effects.

While climate change strongly influences the question of how to save or manage species and the cost and feasibility of doing so, in the end the decision of whether to focus on a particular species still comes down to values.

# Chapter 10

# *The Role of Connectivity*

When we try to pick out anything by itself, we find it hitched to everything else in the Universe.

—*John Muir*

Increasing connectivity is perhaps the most frequently recommended adaptation strategy for maintaining biodiversity in the face of climate change (Heller and Zavaleta 2009). Based on well-documented negative effects of habitat loss and fragmentation on species richness, practitioners have long assumed that the ability of species to move easily across habitats must play an important role in maintaining biodiversity. In the face of climate change, connectivity may become even more important, given its potential to support natural adaptive responses. Connectivity along climatic gradients may facilitate populations' ability to track appropriate climatic conditions, for instance, or allow the flow of genes from warm-adapted populations to those in cooler but warming parts of a species' range.

Connectivity is not without risk. Just as it facilitates the movement of species we humans value, it can facilitate the spread of diseases, parasites, and nonnative species, and can also decrease the ability of populations to adapt evolutionarily to specific local conditions. Such caveats do not negate the importance of considering connectivity, but they do suggest that it needs to be done carefully and with well-designed monitoring in place.

## Why Connectivity Matters

At the most general level, connectivity can promote overall resilience by maintaining robust ecosystems as well as supporting processes and communities important for overall ecosystem health and function. In experimental landscapes, corridors increase species richness as well as the flow of ecological processes such as pollination and seed dispersal. Such genetic mixing among populations can reduce the negative effects of inbreeding and the possible loss of genetic diversity due to the rapid rate of genetic drift in small populations.

Connectivity also supports resilience by facilitating species range shifts. During past periods of dramatic climate change, many species' ranges shifted dramatically, tracking appropriate conditions toward higher latitudes and altitudes during periods of warming and back toward the equator and lower altitudes during periods of cooling. For instance, as the Laurentide Ice Sheet in North America shrank and disappeared following the last glacial period, the latitudinal range of spruce trees shifted far enough north that there is no overlap between their current range and their range 20,000 years ago (fig. 10.1). In contrast, oaks expanded their range such that their current distribution includes both regions they occupied during the glacial period and new regions they have colonized since. Still other species changed their elevational but not their latitudinal range (Davis and Shaw 2001). Such distributional changes are already evident for numerous species in response to current warming. California's Edith's checkerspot and Spain's sooty copper butterflies, for instance, have disappeared from much of the southern parts of their ranges while establishing new populations farther north than before, and an oyster parasite along the eastern coast of North America has expanded its range northward by hundreds of kilometers in just a few years (these and other examples reviewed in Parmesan 2006). In order for these range changes to occur,

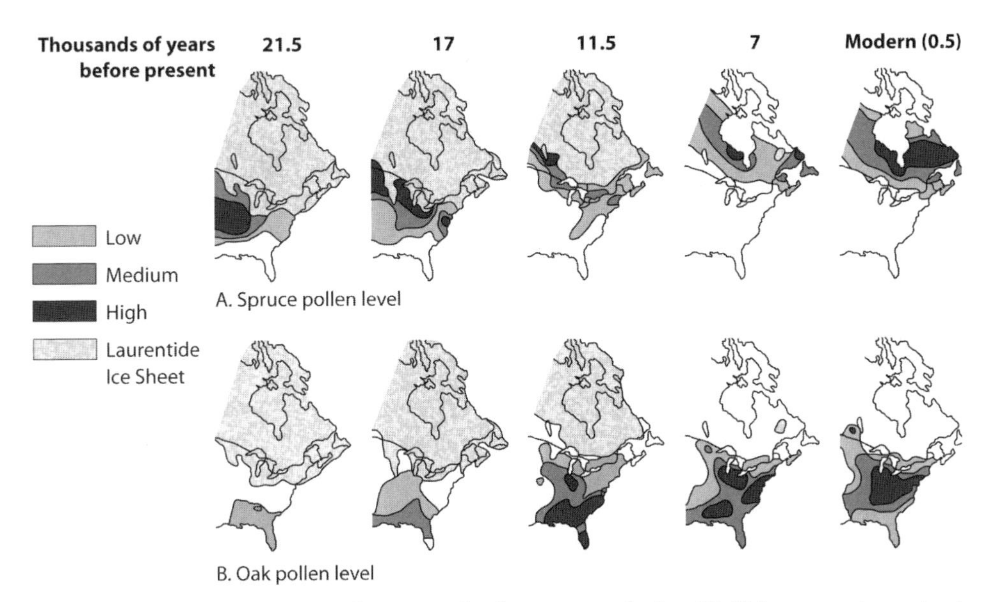


FIGURE 10.1 Range changes of spruce and oak trees over the last 21,500 years as determined by pollen in sediment cores. Pollen percentages indicate relative abundance of each species. After Davis and Shaw 2001.

### BOX 10.1 HOW DO RANGE CHANGES HAPPEN?

One common misperception about species range changes is that they reflect individual organisms picking up and moving when climatic conditions are no longer suitable. Although some animals may be able to respond this way to changing conditions, this is not generally how range changes happen.

All organisms have some mechanism for dispersal, be it passive dispersal by wind or water currents at various stages of life, or flying, walking, swimming, or hitching a ride. Individuals that end up in suitable habitats may flourish and produce offspring, while those that land in unsuitable habitats do not. As the climate changes, individuals may survive in areas where no previous individuals of that species had been able to, or die in areas that once were suitable. Thus do species range changes happen. Climate-related range shifts can also result from altered distribution of key predator or prey species in response to climate change.

species must be able to move across landscapes, seascapes, or freshwater systems—that is, there must be sufficient connectivity.

In addition to facilitating range shifts, connectivity can enhance the ability of populations to undergo evolutionary adaptation to climatic changes by enhancing genetic mixing among populations. This mixing, along with overall genetic diversity, can play a critical role in a population's ability to persist in the face of change by increasing the likelihood that there will be at least a few individuals whose genetic makeup allows them to cope with new climatic situations. There are a number of well-documented cases of populations evolving rapidly in response to other anthropogenic stresses including pollution and introduced pathogens, predators, or hosts (Rice and Emery 2003), and populations have certainly adapted evolutionarily in the past to climate change and variability. Given the rapid rate of current climate change, not all populations will be able to evolve rapidly enough to avoid negative effects of climate change in situ, but supporting the potential for evolutionary adaptation increases the chance that populations and species will successfully respond to climate change through some combination of evolution and range shift. Connectivity supports both these processes, and thus is an opportunity to hedge our bets by maximizing the genetic and geographic options for species and ecosystems responding to climate change.

Finally, connectivity helps to sustain spatially separated populations of the same species, particularly in the aftermath of extreme events. For instance, when a population is decimated by some event such as a cold snap, heat wave, flood, epidemic, or other natural disaster, one element of recovery is recolonization of the site by individuals from outside of the disaster zone. Because climate change is expected to increase the frequency and intensity of several types of extreme weather events (e.g., flooding, wildfires, or droughts), the ability of individuals to move into and recolonize devastated areas is likely to become increasingly important.

**BOX 10.2 DOUBLE JEOPARDY**

Today's species face a twin challenge when it comes to range changes. The current rate of warming is so rapid that it may require range shifts of unprecedented speed, and these changes must take place despite enormous anthropogenic barriers to movement. Barriers include an array of infrastructure and land-use changes such as roads, dams, industrial areas, fences, deforestation, and agricultural, urban, and periurban development. During a past period of rapid warming at the Paleocene/Eocene boundary roughly 56 million years ago, the global average temperature rose 6°C in 20,000 years, leading to the extinction or massive geographic range shift of a number of species. Temperatures in the Northern Hemisphere were warm enough that palm trees, alligators, and other warm-adapted species lived in areas that are today High Arctic tundra. While the magnitude of the current warming will be similar to that at the Paleocene/Eocene boundary, it will occur over just one or two centuries rather than over thousands of years. Furthermore, population declines are likely to limit the ability of many species to adapt evolutionarily to climate change.

## Creating Connectivity Plans for Climate Adaptation

While the concept of connectivity is fairly straightforward, how to make it a reality is not. Different species move in different ways, and what is a barrier for one species or in one ecosystem may not be for another. Although corridors have received the most attention in connectivity literature, they are not the only approach. In marine ecosystems, connectivity is more commonly addressed in the context of ocean currents, geographic complexity, and the duration of free-floating larval stages, and proximity is not always a good predictor of connectedness. In freshwater systems, connectivity is more directional and often must be approached on the catchment level. In addition, water as a medium facilitates connectivity more readily than air in terrestrial systems.

As climate change progresses, the ranges of many species are expected to track preferred climate conditions. Thus corridors or other connectivity plans should be designed to facilitate movement along climate gradients (fig. 10.2). Such gradients often track north-south or low-high gradients, although they can also reflect local or regional topographic and weather patterns. In aquatic systems, connectivity planning should consider the movement of species to water bodies that are deeper, cooler, or less susceptible to seasonal evaporation. For instance, conservation plans could focus on creating or maintaining corridors between reserves along a latitudinal or altitudinal gradient, or on restoring connections between smaller water bodies more vulnerable to warming and larger, more thermally stable ones.

Such an approach is clearly very general, and scientists are working to generate more precise estimates of future range shifts for particular species or entire communities by combining climate, connectivity, and, if appropriate, oceanographic or

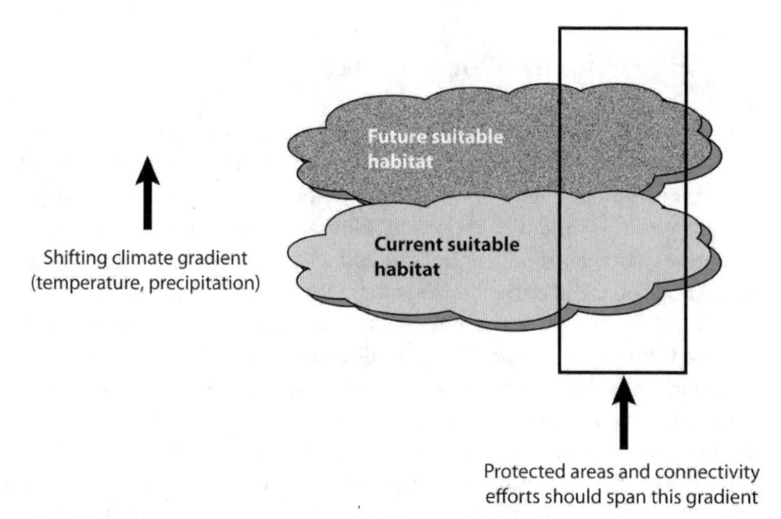


FIGURE 10.2 Designing protected areas to span climatic gradients. Protection schemes that allow individuals and species to move along climatic gradients may help support natural adaptive responses to climate change. While not all species ranges will shift with climate change, many will.

hydrologic models. Although still very much in development, there are a number of promising approaches. In a GIS-based least-cost path analysis, weighting functions are assigned to landscape features such as land cover, population density, distance to major roads, or water. A spatial algorithm then identifies paths across the landscape that should be easiest for the organisms of interest. More recently, some researchers have begun to use circuit or network flow theory, developed to predict electrical conductance in circuit boards, to model gene flow across landscapes (e.g., McRae et al. 2008). In such models, genetic isolation is a function of both distance from other individuals and the "resistance" of the habitat, with flow occurring more rapidly through good, well-connected habitat, and slowly or not at all where habitat is poor or anthropogenic barriers exist. Phillips and coauthors (2008) used a network flow analysis incorporating dispersal ability, climate change projections, and the spatial and temporal availability of habitat for a group of plants in South Africa to design a set of protected areas that maximized long-term persistence while minimizing the total area of new land put under protection. Combined climate and landscape, seascape, or catchment models can be used more broadly to identify areas that may be particularly important for connectivity in a changing climate, targeting them for corridors, broader linkage areas, stepping-stone reserves, or specific use restrictions at particular times of year.

A combination of climate models and empirical data about species' characteristics such as physiological tolerances and dispersal ability can be used to inform decisions about the utility of connectivity strategies as conservation tools for particular groups. For instance, if a species is unlikely to have any remaining climatic niche anywhere under future climate scenarios, increasing connectivity may buy time but in the end will be an insufficient conservation measure. The same would be true for species that are unlikely to be able to shift their range rapidly enough on their own to keep pace with climatic changes. Climate models with a sufficiently fine spatial resolution might be useful in identifying climatically appropriate areas to which such species could be moved, but such efforts are often costly, risky, and ineffective.

## Act Now!

While models such as these can be useful tools for allocating scarce conservation funds, they contain significant uncertainty. As with so many aspects of climate adaptation, we must learn to make good best practice decisions based on current knowledge, adjusting plans as new information or model results become available. What matters is that we do more than chronicle loss and decline while waiting for the right model or the right study to be done. We need to exercise feedback learning to adjust conservation goals based on changing conditions.

## Corridors: For Whom Should We Design Them?

Historically, most corridors have been designed for single species, in particular for large mammalian predators. This single-species focus results in part from the difficulty of designing corridors for groups of species with very different mobility patterns, but also from the premise that designing measures for certain focal species can provide broad benefits for other species.

For some species, particularly large organisms or those with high dispersal capability, corridors may contain only marginal habitat yet be effective conduits for movement between reserves or other suitable habitat patches. In this case, individual members of the species will merely pass through the corridor, as the habitat condition may not be as much of a barrier to their movement. For other species, particularly small and dispersal-limited species, corridors will need to meet all habitat requirements for survival and reproduction. For these species, individuals will not simply move through a corridor on their way from point A to point B; rather, individuals from successive generations will need to be able to disperse gradually through the corridor. Deciding what species to prioritize as targets for connectivity and how to create a permeable landscape for them will take a mix of models, empirical data, and on-the-ground experience.

Designing and creating separate corridors for each species individually clearly takes more time and potentially more political and economic capital than creating multispecies corridors or broader linkages across landscapes. Thus connectivity that is optimized within fully intact landscapes (connectivity by design) is far preferable to connectivity that is a function of remnant contiguous landscapes surrounded by human

## BOX 10.3 ROADS AND DAMS

Much work on connectivity and corridors centers on roads. In addition to being a direct source of mortality for animals, roads are a major impediment to movement across the landscape for smaller or less mobile animals. From a climate change perspective, the role of roads in fragmenting landscapes is made even more important in the United States because many of the major highways stretch east-to-west, hindering northward range shifts. Highway crossings, including underpasses, culverts, and bridges, are an increasingly popular solution. Wildlife crossings across the Trans-Canada highway in Banff are the most carefully studied, and are extremely effective in connecting habitat across the highway. For a period of four years from 2005 to 2009, nearly 84,000 individual wildlife crossings were documented across Banff crossing structures.

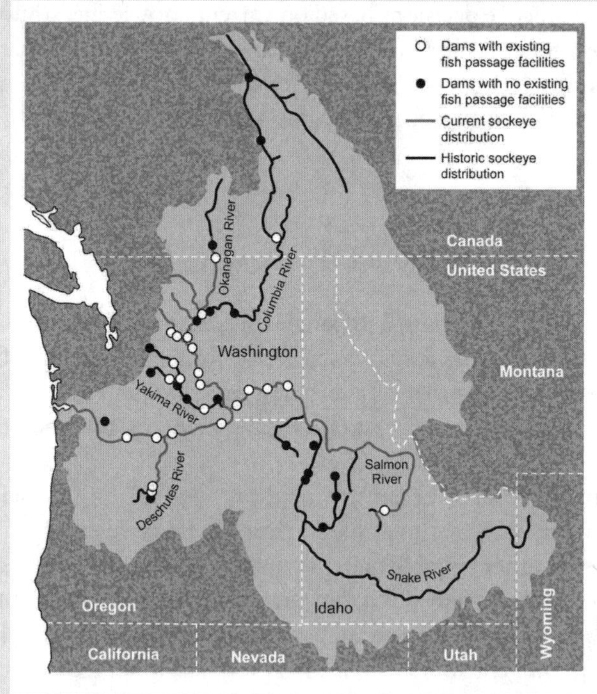


FIGURE 10.3 Historic and current Columbia Basin sockeye salmon distribution in relation to dams and fish passage facilities. After Augerot 2005 and U.S. Army Corps of Engineers 2003.

Dams can present a more daunting challenge. Large dams are an absolute barrier to most riverine species, and most do not have structures to allow the safe transit of fish or other animals past the dam. Forty percent of historical salmon and steelhead habitat in the Columbia River Basin is now inaccessible to these migratory fish because of dams (fig. 10.3). As populations of commercially or culturally important fish have dropped to dangerously low levels, political will is building for greater compromises in terms of water flow rates, bypass structures, and even outright dam removal.

development (connectivity by default). Some practitioners are experimenting with corridor designs that would support multiple focal species and potentially maximize utility across a wide range of taxa, and even existing corridors clearly support many species beyond those for which they were designed. Highway crossings designed for panthers in Florida, for instance, are used by everything from birds to alligators. It is important to recognize that projects designed at the multispecies level may not be as effective for the species involved as individually targeted efforts. Multispecies recovery plans under the Endangered Species Act are not as effective as single-species recovery plans, for instance (e.g., Clark and Harvey 2002).

Because species exist as part of biological communities and interact in a number of ways with other species in their communities, connectivity should also be considered at the level of entire communities rather than single species. Ideally, we will preserve not just lions, elephants, and giraffes, but entire savanna communities. Rayfield and coauthors (2009) have suggested an approach for incorporating consumer-resource dynamics into reserve design, but there is little such work for corridor design.

## Beyond Corridors: Managing the Matrix

For a number of reasons—the aforementioned need to support the movement of entire communities, incomplete understanding of how to create effective corridors, the difficulty of creating sufficient numbers of completely intact corridors, uncertainty about future species ranges, and more—conservation advocates are increasingly considering an approach known as managing the matrix. The matrix refers to landscapes dominated by human land use and development, and softening the matrix means making it easier for species to move through it and restoring its ecological potential for conservation (e.g., fig. 10.4). Again, "moving" may refer to everything from individual animals traveling from one place to another; seeds, embryos, or larvae being carried by air or water currents; or asexual expansion of clonal organisms. Managing the matrix is appealing because species range shifts will occur on regional scales, making regional coordination across political boundaries and agency jurisdictions important to planning. There are a number of approaches being proposed for softening the matrix, including:

- *Riparian corridors*. These are areas along streams and rivers. Creating or restoring wider buffer zones could improve the condition and thus the permeability of the stream or river for aquatic species, while also providing a good corridor for terrestrial species. Because rivers and streams naturally follow an altitudinal gradient, riparian corridors could facilitate upslope movement.
- *Sustainable forestry*. While clear-cutting creates a landscape that is impermeable to most species, it is possible to manage forests and forestry in such a way that working forests continue to provide habitat permeability. Options include a selective harvest regime that maintains a healthy forest at all times, or multispecies forestry.

- *Agroforestry and other agri-environmental schemes*. As with forestry, farms can be managed in ways that make agricultural land a dispersal barrier for most species or in ways that create, restore, or enhance wildlife habitat. The latter approach is widely embraced by smaller organic farms, biodynamic farmers, and others, and can include such measures as planting hedgerows, maintaining well-vegetated riparian areas, or growing crops, such as coffee, in partially forested settings. Some farmers have increased permeability as a result of activities undertaken for other reasons, such as interplanting trees and crops to reduce erosion and drought.
- *Conservation agreements*. Putting agreements in place now with landowners in areas anticipated to be important for species range shifts could make it easier to enact necessary changes later. For instance, owners might sell the development rights to their property.

Determining the optimal use of these and other approaches to connectivity will rely on improved understanding of how species move through various types of habitat,

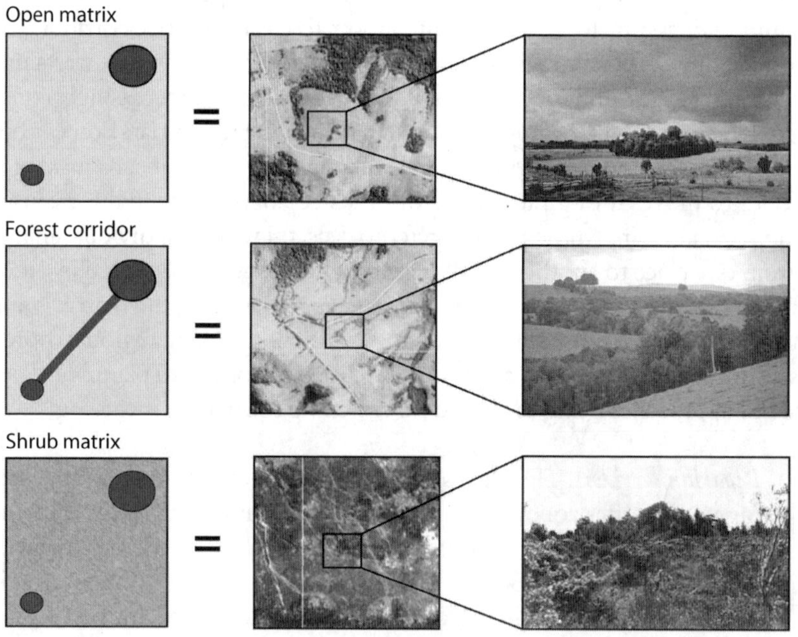


FIGURE 10.4 Experimental treatments testing the effect of habitat type on the dispersal of Chucao Tapaculos (a species of bird) in Chile's temperate rainforests. All birds released into a forest patch surrounded by shrub matrix and most released in areas forest wooded corridors dispersed successfully within thirty days, whereas only half of those released into patches surrounded by open matrix dispersed. After Castellón and Sieving 2006.

and both models and monitoring will be important. Studies may focus on particular places, as when a species is known to travel along a particular route (e.g., mating frogs or migrating grey whales), or on particular types of habitat modification. Currently, there are many efforts to soften the matrix in areas adjacent to reserves or protected areas. Generally softening the matrix, like generally increasing population resilience, may be more appropriate than complex, model-driven approaches in situations where knowledge and resources are limited. A more targeted approach would be to focus softening efforts along climate gradients that link reserves in warmer zones to reserves in cooler areas. Such gradients often follow elevation or latitudinal gradients, but they may also reflect local or regional topography and weather patterns. Managers at Joshua Tree National Park in California are working with surrounding communities to maintain and establish corridors and open space near the park. The village of Joshua Tree is working specifically to connect Joshua Tree National Park with Bureau of Land Management land to the north. Such efforts also represent a "no regrets" strategy. Even if softening the matrix around a reserve does not end up having a significant effect on connectivity, it still increases the effective reserve area by minimizing edge effects.

Managing the matrix is in essence a form of adaptive management. There is much uncertainty about future climate and how species and communities will respond to it. Managing the matrix and getting people to think about the matrix as a whole keeps more options open as these changes manifest.

## Thinking Big

Some countries have established large, coordinated connectivity projects within their own boundaries, such as Australia's 2,800-kilometer Great Eastern Ranges Initiative (formerly Alps to Atherton) or Great Barrier Reef Marine Park. There are also a growing number of transboundary efforts. One of the most established is the Yellowstone to Yukon (Y2Y) Conservation Initiative, which began in 1993 as the brainchild of a group of scientists and conservation practitioners. Running through five American states, two Canadian provinces, and two Canadian territories, it represents one of the first efforts to put the principles of large-landscape conservation into practice across a mountain environment, and today encompasses a 3,200-kilometer north-south corridor ranging in width from 500 to 800 kilometers with more than 700 protected areas and covering a total area of 1.3 million square kilometers.

The corridor's immense latitudinal and elevation range is ideal for promoting species and community adaptation to climate change. It will facilitate the poleward and upslope range shifts expected to play an important role in persistence in a warming world. Rather than seeking to turn an entire region into a protected area, the goal is to work with the entire range of stakeholders in the region—private and public landowners, businesses, tribes, hunters and anglers, environmentalists, and more—to manage the area as a whole in a way that promotes connectivity and conservation while supporting multiple uses.

## Final Thoughts

Increasing connectivity clearly has potential as an important element of adapting to climate change. At the very least, it is likely to increase the overall resilience of populations, communities, and ecosystems, and it may reduce vulnerability to threats posed specifically by climate change, particularly if focused along climatic gradients. Increasing connectivity is not, and in fact cannot be, about simply turning as much land and water as possible into protected areas or corridors. We will have to work with diverse stakeholders to manage land as a patchwork that maximizes permeability. Similarly, range expansion and genetic exchange do not stop at political boundaries: there needs to be regional coordination across political boundaries and agency jurisdictions.

Despite the intellectual appeal of connectivity and some hopeful examples, there is still limited evidence for corridor effectiveness, minimal understanding of how to optimize connectivity across landscapes in practice, and very little knowledge about how to minimize risks of increased disease and invasive species transmission. Because so much of what we believe about corridors and connectivity thus far rests primarily on intuition and ecological theory, it is essential that efforts at implementing them take place in the context of adaptive management. Monitoring plans must be an integral part of such efforts.

# Chapter 11

# *Restoring for the Future*

In the long run, no inherent natural ecosystem or landscape configuration exists for any region.

—*Stephen Jackson and Richard Hobbs*

By its very nature, restoration requires making choices. These may be choices about where to restore, how to restore, or what to restore to—climate change has implications for all of them.

A common goal of restoration has been to attempt "the return of an ecosystem to a close approximation of its condition prior to disturbance" (National Research Council 1992). Predisturbance or reference condition was defined by some historic state or, more recently, by the historic range of variability. The latter approach (known as HRV) recognizes the inherent complexity and dynamic nature of natural systems, and has led to greater emphasis on restoring ecological processes rather than simply the static characteristics of a location such as tree density or patch size. Climate change and invasive species, however, call into question whether even HRV will remain a valid approach: it is likely that the historic range of variability will no longer be possible for most ecosystems. In the immortal words of Chris Milly, "stationarity is dead" (Milly 2007).

How, then, might the goals and tools of restoration be adjusted to better reflect our rapidly changing world? New goals could focus on restoring sites in ways that attempt to decrease the rate or extent of change; maintaining the ecosystems' services, if not the particular species assemblages that currently exist; or facilitating a shift of community structure and function in anticipation of future climate conditions.

**BOX 11.1 CLIMATE CHANGE AND DUNE RESTORATION**

In 1954 and 1986, the Netherlands initiated dune restoration projects relying primarily on sod removal. These projects led to the reappearance of target pioneer vegetation within a few years, a clear success. Similar attempts in 1990 and 1995, when precipitation levels were significantly higher, had completely different results. Target species either never reappeared, or reappeared only briefly, and species adapted to wet, eutrophic conditions took over instead. Efforts to restore marshes in the same region over the same time period saw similar results. One cannot assume that a single approach will work in all climate conditions, or that similar objectives are appropriate to all climate conditions, stable or changing (based on Choi et al. 2008).

This latter option highlights some philosophical quandaries. Does facilitating a shift toward future conditions mean giving up altogether on particular species? If not, what is the best option for species or populations projected to lose most or all of their range? For instance, climate projections suggest that many river reaches in California's Tahoe National Forest currently targeted for salmon restoration may become climatically unsuitable for salmon within twenty years regardless of restoration efforts, leading managers there to consider whether restoration for salmonids is a reasonable expenditure of effort (Julius et al. 2008).

On a more positive note, the fact that many restoration projects by their very nature involve intervention in the biotic and abiotic elements of a particular location means that they provide valuable opportunities to build features that decrease vulnerability to climate change into the fabric of the system. They also allow for a greater range of experimentation with adaptation approaches that can then feed into adaptive management decisions.

## Restoration Strategies to Reduce Change

Although individual restoration projects cannot by themselves stop global climate change, they *can* influence local climate, reduce greenhouse gas emissions, or reduce the negative effects of global climate change. Thus one target of restoration projects might be "climate restoration," maximizing the ability of local or regional ecosystems to maintain historical climate patterns or at least to slow the rate of change. Protecting or restoring lowland forests as part of projects to restore or conserve downwind montane cloud forests would be an example of this type of approach (Ray et al. 2006). Alternatively, projects might target ecosystem functions likely to be affected by or become more important because of climate change, such as surface or groundwater storage, floodwater dissipation, or erosion control.

*Support Optimal Coastal Sediment Input and Retention*

On a local scale, the rate of sea-level rise reflects not just global changes and regional tectonics, but the rate of coastal accretion or erosion as well. Thus increasing sediment input and retention can in some cases decrease the rate of relative sea-level rise locally. Excess sediment input can be problematic, however, as seen in massive mangrove die-offs from sedimentation following heavy flooding events linked to the 1997–1998 El Niño in East Africa (Kitheka et al. 2002), or the death of juvenile sea turtles unable to dig their way out of nests following heavy flooding and sedimentation in eastern Madagascar (B. Randriamanantsoa, pers. comm.). Restoration efforts, then, must seek a balance between too much and not enough.

Coastal sediment sources include bluff and cliff erosion, rivers and streams, alongshore transport, and deposition and decay of organic matter. Because the relative importance of various sediment sources varies significantly from location to location and will also change with climatic changes, coastal restoration projects addressing sediment supply issues must assess the local reality. In the Santa Barbara littoral cell along the coast of southern California, for instance, rivers historically account for more than 99 percent of sediment input, with bluff erosion supplying the remainder (California Department of Boating and Waterways and State Coastal Conservancy 2002). The nearby Oceanside littoral cell (see relative location in fig. 11.1) may receive more than 60 percent of its beach material from bluffs, with the rest split evenly between input from rivers and gullies (Young and Ashford 2006). In the former case, removal of shoreline armoring would have little or no effect on restoring natural beach sediment supply, while in the latter it could play a significant role.

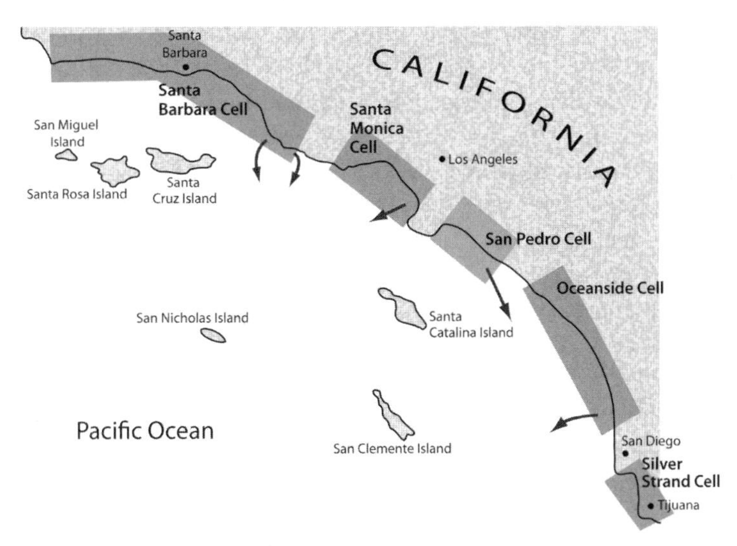


FIGURE 11.1 Map of littoral cells in Southern California. A littoral cell is a distinct section of coastline with self-contained sediment transport cycles (i.e., sediments cycle primarily within that segment of coast).

In addition to removing shoreline armoring, restoring beach sediment supply may involve removing dams, bypassing sediments around dams, diverting some portion of river flow, reversing stream channelization, limiting sand mining, removing jetties, and restoring watersheds. For example, the West Bay Sediment Diversion Project in Louisiana will create a diversion channel in the west bank of the Mississippi River in southern Louisiana to restore input of the Mississippi's sediment-laden water to one section of the Mississippi delta, with the expectation of restoring more than 10,000 acres of marsh over a twenty-year period. The California Beach Restoration Study suggests a protocol for identifying and prioritizing reservoirs and debris basins for sediment supply intervention projects aimed at beach restoration (California Department of Boating and Waterways and California Coastal Conservancy 2002).

Sand and sediment addition can play a role in shoreline restoration and maintenance, but cost and limited sediment supply often make this a challenging approach over the long term. Indeed, many shoreline nourishment projects have insufficient planning or monitoring to even evaluate long-term sustainability or costs (e.g., Hamm et al. 2002; Hanson et al. 2002). Shoreline nourishment may be somewhat more feasible in areas near harbors or channels that are regularly dredged to support commercial use. In such cases, the repeated addition of dredged material may be considered as one element of a long-term strategy in the face of sea-level rise, provided the dredged material does not contain excessive amounts of toxic material. An example of this approach is the Poplar Island Environmental Restoration Project in Chesapeake Bay. Between 1847 and 1990, the island shrank from more than 1,000 acres in size to fewer than 10 acres due to relative sea-level rise (the combined effects of global changes and local subsidence). In 1994, the Army Corps of Engineers was looking for a beneficial use for excess material from its regular dredging of navigation channels in the bay, and with a coalition of state and federal agencies developed a plan for restoring Poplar Island. Work began in 1998 with the construction of a series of dikes outlining the area to be restored, and by 2005 five of the six priority waterbird species targeted by the project had recolonized the island. By 2013, it is expected that the island will again be larger than 1,000 acres in size (Erwin et al. 2007).

Sediment retention efforts, such as creating groyne fields, replanting or restoring submerged aquatic vegetation, and restoring marshes and mussel and oyster beds, can also play an important role in reducing coastal vulnerability to sea-level rise. Creating or restoring vegetative or dune buffers, important restoration projects in their own right, gain added importance if used to protect important habitats or restoration sites on the landward side.

The flip side of the sediment equation is the need to prevent damaging sediment deposition during high-flow events. This may involve linking coastal restoration efforts with programs higher in the watershed, such as restoring upland vegetation to limit erosion during heavy rains. Creating side-channel basins or structures that facilitate overflow into floodplains during high-water events can also help to retain more sediment upriver.

### *Support Climate-Savvy Hydrology and Geomorphology*

Restoration projects commonly address issues of disrupted or altered hydrology and related geomorphology, such as channelization or dam-building. Such projects must clearly take climate variability and change into account. For example, restoration projects in areas predicted to experience increased frequency and intensity of flooding might put increased focus on interventions such as off-channel storage basins, regrading stream banks to allow natural bank overflow into floodplains during high-water events, restoring meanders or woody debris to slow flow, or removing impervious channel linings that limit erosion and water absorption.

In flood-prone areas, appropriate grading or terracing of slopes may be particularly important during periods of revegetation. The value of terracing for limiting erosion has been repeatedly demonstrated in regions as diverse as the high Andes, Malaysia, and Central America. In coastal areas that may experience increasingly high storm surges, it may likewise be important to create grades that will facilitate runoff of the salty water from storm surges, and to eliminate drainage ditches that could allow more rapid saltwater intrusion into freshwater areas (see Box 11.3).

### *Create or Restore Microhabitats and Refugia*

The existence of a climatically complex landscape can decrease vulnerability to climate change by allowing flora and fauna with a diversity of climatic needs to coexist, and by

#### BOX 11.2 THE PAST, PRESENT, AND FUTURE OF MONO LAKE

Mono Lake, in California's Great Basin, is a highly saline basin lake with no outlet. Following the diversion of freshwater from Mono Lake's tributaries by the Los Angeles Department of Water and Power in 1940, lake level dropped rapidly, salinity increased, and lake biodiversity and productivity declined. As part of a court case to set acceptable water levels for Mono Lake, scientists studied 4,000 years of Mono Lake's hydrology and geomorphology. They identified a critical water-level threshold below which significant shifts in the lake's ecosystem, hydrology, and geomorphology were likely to take place. They also noted that periods of extreme drought leading to perilously low water levels such as those that occurred during the Medieval Warm Period are not improbable for California's near-term future, suggesting that a significant buffer was needed to ensure that lake levels did not fall below the threshold. As a result the court decided to realign lake levels based on historical and projected future conditions rather than simply mandating a restoration to pre-diversion levels. Water levels at the time diversion began were anomalously low after the Dust Bowl decades of the 1920s and 1930s (Millar and Woolfenden 1999).

providing climate refugia to which animals may retreat during periods of particularly high climate stress. Restoration projects can be designed to create complexity and climate refugia within project areas themselves, or project locations can be selected to create refugia or climatic complexity within a broader landscape. For instance, restoration efforts may target locations that increase the frequency of coldwater patches in streams by influencing surface tributary or groundwater inflow (Watanabe et al. 2005), a potentially important factor for the persistence of coldwater fishes. Restoration of larger landscapes, such as slag heaps, might aim to retain varied topography, creating shaded ravines, sunny slopes, and planting with a range of species suited to a variety of microhabitats.

## Restoration Strategies to Support Resilience

In addition to reducing the rate and extent of physical and chemical changes, restoration projects can be selected and designed to maximize ecosystem resilience to unavoidable change. All of the usual elements of restoration planning apply, but adding a climate change lens may alter tactics and priorities. Restoration projects that would have the added benefit of increasing ecosystem resilience or supporting the resilience of species of concern may be given increased priority. An example would be restoring mangrove forests near coral reefs. In addition to creating healthy mangrove communities, such projects could reduce the vulnerability of corals to bleaching in several ways. Mangrove forests release dissolved organic matter, which provides a partial sunscreen for corals and thus reduces risk of bleaching. By providing nursery habitat for herbivorous fish, mangroves might increase a reef's likelihood of recovering from bleaching by increasing populations of adult herbivores. And both healthy mangroves and healthy reefs can be critical for decreasing the vulnerability of inland areas to erosion and storm surge.

### *Anticipating Species Shifts in Response to Change*

One of the most straightforward approaches to decreasing the vulnerability of restoration projects to climate change is to anticipate the movement of individuals and populations to track appropriate conditions. This includes the inland movement of coastal systems as a result of sea-level rise, the upslope movement of many terrestrial systems in response to warming temperatures, and the poleward movement of both terrestrial and aquatic systems. Put another way, we should protect and restore future as well as current habitat, and the transition between the two. Such transitions may be gradual or abrupt. For instance, shorelines currently protected by barrier islands may be rapidly and radically reshaped should those islands be breached or disappear (fig. 11.2).

There are a number of ways this could play out in practice. For instance, coastal wetland mitigation projects might be required to cover an elevational range such that the restored or created habitat includes both areas appropriate for wetlands today, and

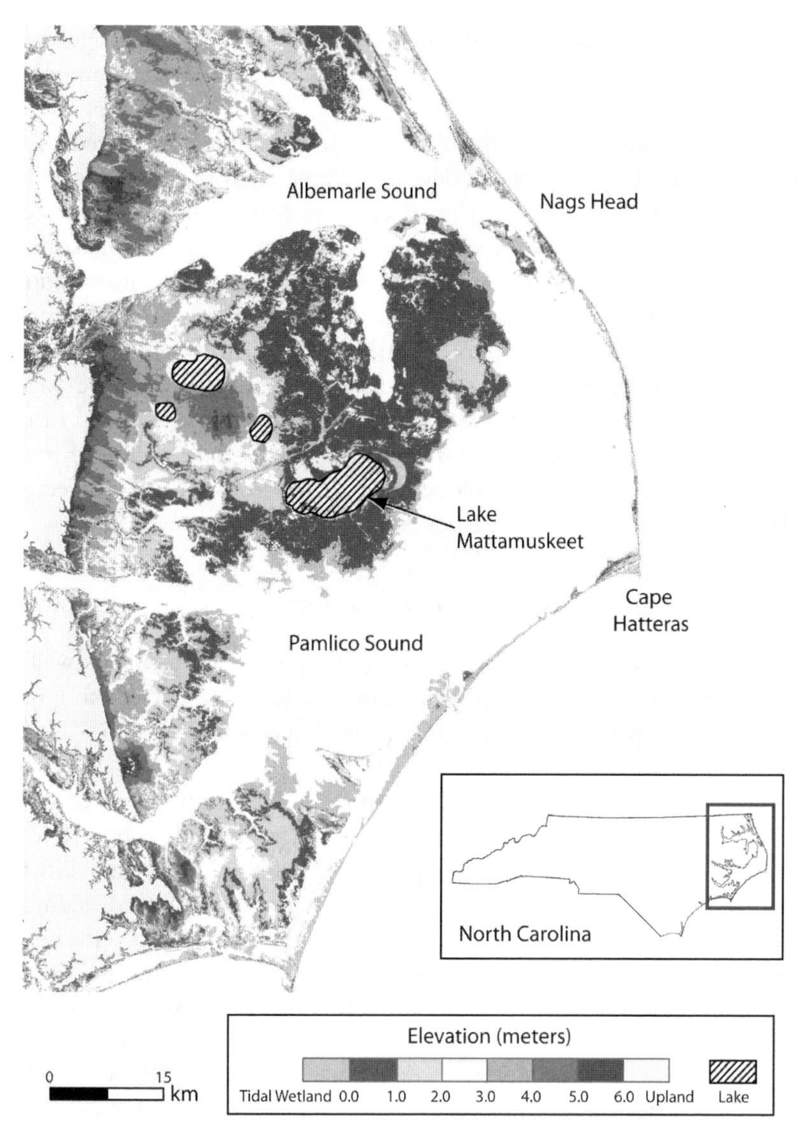


FIGURE 11.2 Map of the Albemarle/Pamlico Sound region of North Carolina. Lake Mattamuskeet—the state's largest natural lake—will be lost during the next century under even low-end sea-level rise projections. After Titus and Wang 2008.

those that will be appropriate for wetlands in fifty years. Topography can be designed to facilitate water runoff after storms or high tides, promoting plant growth and marsh expansion, and to support a slow enough transition from fresh to brackish to salty conditions that plant communities will be able to adjust. Wetland restoration funds might be used to secure and restore areas that are not currently wetlands but will become wetlands as sea-level rise progresses. Likewise, funds earmarked for the restoration of prairies or other particular habitat types might be used to restore or manage areas poleward of existing target habitats to facilitate the successful shift of the community as climate change progresses.

Restoration and acquisition prioritization criteria could be updated to include consideration of local topography and development patterns that facilitate inland, upslope, or poleward habitat shifts. For instance, it may make more sense to restore a coastal wetland backed by gentle slopes with little development rather than one backed by steep cliffs, since the latter will inevitably be lost as sea level rises. Restoration plans could include the removal of barriers to habitat shifts, such as shoreline armoring or logging roads, even when they fall outside of the project site. Where outright removal of barriers is not possible, other options for increasing connectivity may be considered, such as wildlife highway crossings (see chapter 10 for further discussion of connectivity). Restoration projects may also take steps to prevent the creation of barriers to habitat shifts within a set distance upslope, poleward, or inland of the project site, for instance by securing conservation easements from relevant landowners.

### *Restore with a Mix of Species*

Just as practitioners should consider restoring and protecting future as well as current habitat, their choice of flora and fauna should reflect both current and future expectations for variables such as salinity, pH, flood frequency and severity, or drought frequency and severity.

If uncertainty about future conditions is not too high in terms of either the direction or timing of change, practitioners may take a deterministic approach to species selection. In this case, species should be selected based on their likely suitability for projected future conditions. Thus the suite of species or varieties planted might include shorter-lived species that do well under current conditions, species with intermediate life spans that do well across a range of conditions, and longer-lived species better adapted to future conditions. Mixing species with different life spans, times to maturity, and climatic needs may facilitate community adjustment to changing conditions over time, and short-generation plants adapted to current conditions may facilitate the establishment of longer-lived plants adapted to expected future conditions.

If the rate, extent, or direction of change is uncertain or the prediction is simply for more climatic variability, a bet-hedging strategy may be more appropriate. In this case, practitioners should select a mix of species of all life spans adapted to a range of conditions. This increases the likelihood that some species will do well in whatever future comes along, and also provides a diverse seedbank to allow shifting species composition under variable conditions. A gravel pit revegetation project in Colorado took just such an approach. Practitioners planted a mix of short-, medium-, and tall-grass prairie species with a wide range of moisture demands. There was one wet year after the planting, followed by three years of drought. The end result was a mixed-species drought-tolerant community that appeared resistant to invasion by other species (Seastedt et al. 2008).

### BOX 11.3 THE ALBEMARLE PENINSULA

Some of the very characteristics that give North Carolina's Albemarle Peninsula its rich array of swamp forests, marshes, and pocosin bogs make it highly vulnerable to climate change (fig. 11.2). Significant local land subsidence means sea level is rising here at more than twice the average global rate, and the low, flat topography means every inch of sea-level rise causes the shoreline to shift inland by several feet. Federal, state, and private landowners who manage the 400,000 acres of the peninsula currently in conservation are developing plans to lessen the impact of sea-level rise here. Approaches under consideration fall roughly into two categories: conservation and restoration. Acquiring and preserving more land, particularly inland from existing conservation lands, will allow ecosystems to shift landward with the rising seas and allow managers greater flexibility to respond to changes as they develop. Restoration efforts would include a variety of tactics. One critical target is the numerous drainage ditches and canals dug to drain wetlands for farming or mosquito control, which facilitate a more rapid intrusion of saltwater farther inland. Because the peat soil characteristic of this area disintegrates rapidly in the presence of salt, saltwater intrusion into peaty soils would increase the rate of soil loss and thus increase the rate at which land is lost to the waters of the Sound. The Nature Conservancy and partner organizations will use hydrologic models to prioritize areas for actions such as filling the ditches or installing water control structures to minimize saltwater intrusion, as well as experimenting with ways of using water control structures to support enhanced peat growth. A second strategy is to restore wetlands using salt-tolerant plants in areas likely to be inundated. While native bald cypress trees are salt-sensitive, there is evidence that some individuals may be able to survive for decades or centuries following salinization (Yanosky et al. 1995), and scientists are working to identify more salt-tolerant bald cypress trees (Conner and Inabinette 2005) . Finally, restoration of submerged aquatic vegetation and oyster reefs may help to reduce wave energy and erosion or even enhance shoreline accretion.

### *Restoring Ecosystem Engineers and Keystone Species*

Some species exert a powerful influence on the structure of the ecosystems in which they live, either through predation (keystone species) or by physically restructuring the "distribution, abundance, and composition of energy and materials in the abiotic environment" (ecosystem engineers; Jones and Gutiérrez 2007). Both can be used to decrease the vulnerability of ecosystems to climate change.

Loss of wetlands throughout North America is a major environmental problem, and is likely to become even more of an issue as climate change leads to increasingly dry summers across much of the continent. One contributor to wetland loss has undoubtedly been the extirpation of American beavers throughout much of their range, given beavers' well-recognized ability to create and maintain wetlands at landscape scales.

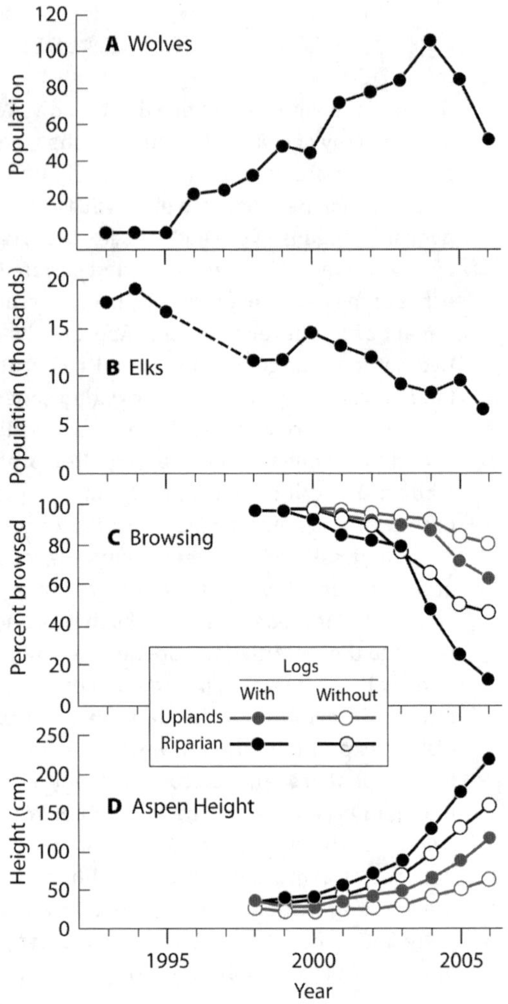

FIGURE 11.3 The effect of wolf predation on elk and aspen populations in Yellowstone National Park. As wolf populations increased following their reintroduction in 1995 (a), elk populations declined (b), which decreased elk browsing on aspen (c) and allowed aspen to grow taller (d). The decreased browsing and increased aspen height were greatest in areas where elk were most vulnerable to predation: riparian areas with downed logs. From Ripple and Beschta 2007.

Recent research suggests that restoring beaver populations may be a critical link not only in increasing wetland size and number, but in reducing their vulnerability to climate change as well. For instance, over a fifty-four-year period, the vast majority of the variability in wetland presence or absence in a region in western Canada was explained by the presence or absence of beavers rather than by climatic variables (Hood and Bayley 2008). During droughts, beaver presence was linked with a ninefold increase in available open water. Thus one option for climate-savvy wetland restoration on a broad and sustainable scale is to reintroduce beavers and generate public support for their presence. There may be an initial need to create or restore some degree of wetland habitat and vegetation in heavily degraded habitats, but reintroducing beavers provides for much more effective and affordable long-term maintenance of wetland habitat. Restoration efforts geared toward increasing or protecting populations of wetland-dependent species such as amphibians and many birds could take place in a broader context of beaver reintroduction. Because climate change is likely to increase length and frequency of droughts in many regions, beaver reintroduction provides a certain level of "wetland insurance."

Wolves are another species that was once widespread in North America but was extirpated throughout much of its original range by hunting and trapping. The loss of wolves has led to a sharp increase in populations of grazers, notably ungulates such as moose, deer, and elk. This in turn has led to a decline in vegetation, which can be particularly damaging along riparian corridors. Wolf reintroduction programs have demonstrated that the presence of wolves supports vegetation recovery (fig. 11.3), which could reduce vulnerability to climate change in a number of ways, including decreased erosion during storm events, increased water flow and availability during dry periods, and cooler in-stream temperatures in summer. Wolves are also likely to decrease the vulnerability of numerous scavenger species to climate change in areas that will see strong declines in winter snowfall. The shorter, warmer winters predicted as a result of climate change will reduce the number of ungulates weakened or killed by deep snowfall, which means less winter food for scavengers. Because wolves typically do a poor job of cleaning the carcasses of their kill, however, the presence of wolves ensures a relatively steady supply of food for scavengers even in low-snow years (Wilmers and Getz 2005).

Yellowstone National Park in the United States provides a strong example of the importance of both wolves and beavers for ecosystem function, as well as the need for multiple levels of restoration. The loss of wolves in the region led to an increase in elk, which dramatically reduced woody forage for beavers in riparian corridors. This led to a striking decline of resident beavers in the park, and a shift from a beaver-willow mosaic landscape to an elk-grassland one. While wolf reintroduction controlled elk populations, it did not lead to a rapid return to the beaver-willow mosaic. The reestablishment of willows has been slowed by the loss of the rich beaver-pond sediment that promotes willow establishment, and by changes in the water table due to the loss of beavers. Without enough willows in the riparian zone, there is not enough food to support resident beaver populations. Creating artificial ponds may help to create willow habitat, and may thus be an important additional step in bringing back resident beaver populations and the resilience to climate change that they provide to the park (Wolf et al. 2007).

### *Promoting Microevolutionary Adaptation*

Although evolution is central to modern biology, it has been strikingly absent from the discussion of conservation and natural-resource management in the face of climate change. Yet there is ample evidence that natural selection has not uncommonly resulted in adaptive evolution of wild populations in response to anthropogenic stressors. One of the classic examples in evolution textbooks, for instance, is the evolution of peppered moth populations in response to increased soot. As soot darkened the bark of birch trees, the formerly dominant white morph gave way to dark morphs. Several wild populations of plants and worms have evolved heavy metal tolerance following increased heavy metal pollution from mine tailings. Hawaiian honeycreepers had bill shapes that maximized feeding efficiency on a particular flower; when that species went extinct, the honeycreepers shifted their feeding to a different species and subsequently evolved shorter bills that allowed better access to this new nectar source. These and other evolutionary responses to human-induced environmental change are reviewed by Rice and Emery (2003).

Current restoration guidelines generally recommend using local seed sources, but this approach may limit opportunities for adaptive biological evolution in the restored populations for several reasons. Starting with a genetically restricted population can create a founder effect that limits the genetic diversity and thus the evolutionary options of the population for generations to come. Also, while local populations may indeed be well adapted to local conditions, those conditions are likely to change as global warming progresses. Increasing the genetic variation in restored populations to include genotypes better adapted to predicted future conditions increases the chance that restored populations will be able to successfully track changing conditions through evolutionary adaptation. The key is to balance the need for genetic diversity with the importance of not swamping unique local or regional genotypes. One option would be to create regional seed mixes delineated by climate zones as is currently the practice, but to include genotypes from a variety of microenvironments within each zone as well as from the edges of each climate zone.

This leads to another possible role for microevolution in making climate-savvy decisions about where and when to support restoration. While the reality of species range shifts has led some managers to suggest focusing limited conservation and restoration funds on the cooler edge of species' ranges (where populations are likely to increase) rather than the warmer edge (where populations are likely to decline), more warm-adapted populations may provide a critical pool of genotypes better suited to future climatic conditions. Giving up on them may remove key genetic options for populations toward the center of a species' range.

## Final Thoughts

Because restoration is inherently more interventionist than many other forms of conservation, it provides an excellent opportunity for experimenting with concepts of resistance and resilience to climate change. An ecosystem must be "broken" to be targeted for restoration, and in the process of "fixing" it we can make it more robust for future climates than it may have been originally. Trying new, climate-savvy approaches to restoration may seem risky, but there is risk in all restoration projects. Also, the long-term monitoring that is part of any well-conceived restoration project makes such work well-suited for the sort of adaptive management we need to create a more robust, responsive approach to a changing climate.

Chapter 12

# *The Hordes at the Gates*

## Beating Back Invasive Species, Pests, and Diseases

Climate change complicates everything.
—*Ecological Society of America*

Climate change will influence invasive, pathogenic, and parasitic species in a host of ways, as it does all species. It will influence their spread and their harmfulness, as well as the success of our efforts to control them, in some cases for better and in others for worse. This chapter investigates the interplay among climate change, pests, diseases, and invasive species, and explores considerations for limiting our vulnerability to all.

### Nonnative Invasive Species

Nonnative invasive species are a major and growing driver of environmental change, costing billions of dollars annually in agricultural, forestry, and public health impacts. They can outcompete native species for resources, alter habitat structure and disturbance regimes, affect plant and wildlife health, and influence human health, recreation, and industry (e.g., Strayer 2010). Some introductions are intentional, such as the use of nonnative species for aquaculture, agriculture, or ornamental plantings, while others are not, such as marine species carried in the ballast water of commercial ships.

Although the rate of invasions has been increasing dramatically (fig. 12.1) and speculation about combined effects of climate change and species invasions abounds, there is relatively little direct demonstration of systematic effects. Reviews of the topic

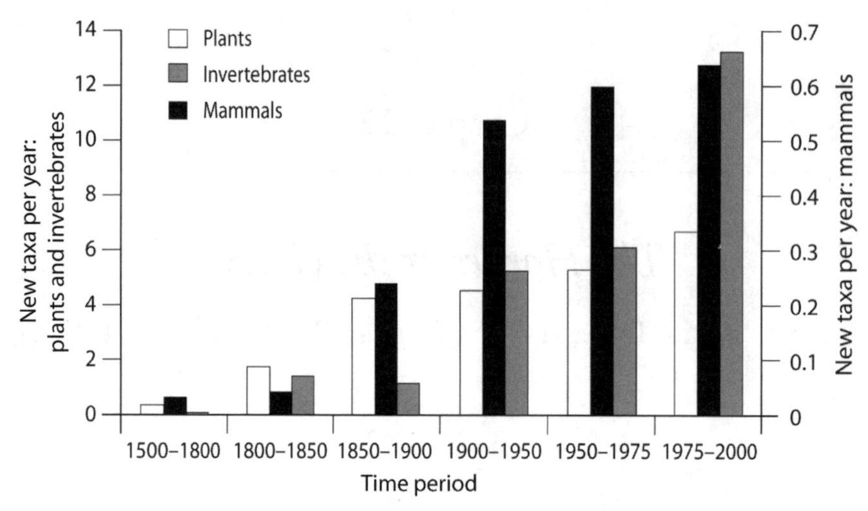


FIGURE 12.1 The number of nonnative species becoming established in Europe since the 1500s. After Hulme 2009.

are largely conceptual and built on what is thus far a small number of studies. Nonetheless, existing examples and theory shed light on what to look out for and potential approaches to decreasing vulnerability to noxious invaders in a rapidly changing climate.

Conceptually, species invasions can be broken into four stages: drivers (how species get to new areas), establishment (whether new arrivals are able to survive and spread), effects (how new arrivals affect communities or ecosystems), and response (how humans respond to nonnatives). Climate change has implications for all stages.

### *Changing Drivers*

By changing patterns of air and water flow, climate change will alter the natural movement of both native and nonnative species around the globe. Many marine

> **BOX 12.1 WHAT'S IN A NAME?**
>
> Species that have been introduced beyond their native ranges by direct human transport (purposeful or not) are variously termed nonnative, alien, exotic, naturalized, introduced, adventive, or invasive in the literature. Some people use the term invasive to describe aggressively spreading species regardless of origin, and point out that nonnative species that are not invasive are not problematic. Changing environmental conditions may allow previously noninvasive species to become invasive, however, and will also lead more species to change their ranges without human intervention.

biogeographic boundaries, for instance, result from barriers to dispersal rather than climatic conditions per se. A classic example is Point Conception, California, where the convergence of north- and southbound currents combines with coastal upwelling to create a "leaky barrier" to movement of species with waterborne propagules around the point (Wares et al. 2001). Changes in current and upwelling patterns in the region, such as the persistence of more El Niño–like conditions, could lead to a northward "invasion" by southern species (Hohenlohe 2004), as could the faster larval development times likely to result from warmer water (Byers and Pringle 2006). Likewise, changes in the Polar Front Zone where Antarctic and sub-Antarctic water bodies meet may contribute to observed and future species incursions into Antarctic (Barnes et al. 2006), and unusual air mass movements in 2000 brought large numbers of a nonnative moth to the Arctic island of Spitsbergen (Coulson et al. 2002).

Climate change is also likely to influence human activities responsible for many invasions, potentially increasing the opportunities for invasion by changing origins, destinations, speed, and volume of transport. One clear example is that decreasing Arctic sea ice extent will open shipping routes not previously accessible (fig. 12.2), thereby opening up new areas to exposure to nonnative species through ballast water, hull-

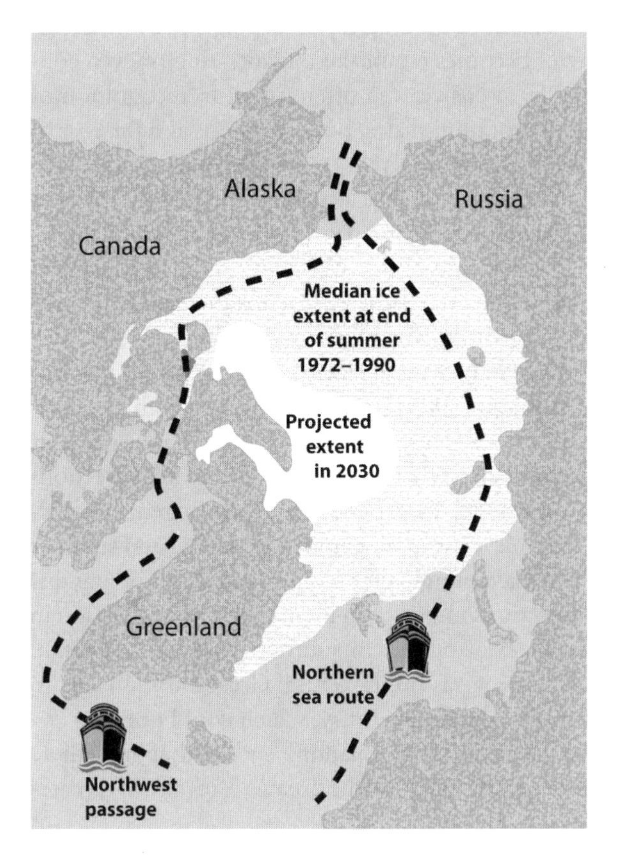


FIGURE 12.2 Projected summer sea ice extent in the Arctic by 2030 and possible trans-Arctic shipping routes resulting from ice loss. Based on maps from the University of Washington's Polar Science Center.

fouling organisms, and hitchhikers in cargo. Increased Arctic shipping could speed up the rate at which invaders travel across this region. Warmer climates around the globe could also allow ornamental or commercial species to be shipped and grown in new locations, expanding introductions of not just those species but their hitchhikers as well.

### *Invasion Success*

Regardless of mode of transport, climate change will influence a species' ability to survive and succeed once it arrives in a new location. The degree to which an invader's success is increased or decreased by climate change depends on the species in question and what currently limits its establishment and spread. For instance, one nonnative olive crop pest in Crete responds positively to warmer conditions, and climate change will allow it to expand into higher elevations(Ross et al. 2008). In contrast, the spread of Argentine ants in Jasper Ridge Biological Preserve in northern California appears limited by low summer rainfall, so the drier summers projected for most of California under climate change should help to limit their spread (Heller et al. 2008).

Identifying regions where climate change may increase the risk of successful invasions would allow resource managers and conservation practitioners in those regions to be particularly alert for new arrivals, and to develop rapid response plans *before* invasions occur. Once an invasive species becomes established, eradicating or even controlling it is generally time-consuming, expensive, and often futile. In recognition of the value of early detection and response for minimizing both invasions and the cost of controlling them, some jurisdictions are taking proactive steps to address invasions. In the United States, Kansas state officials share information about changes in invasive species and possible responses with their counterparts to the north and south. In response to concerns that warmer winters will increase the overwinter survival and spread of the invasive water hyacinth, Arkansas added the species to its noxious weed and prohibited plant lists.

Climate envelope models can be a useful tool for exploring changes in invasion risk under climate change. Bradley and coauthors (2009) used this approach with five invasive plant species in the western United States. They explored which climatic factors, if any, currently limited the ranges of these species, then used regional climate projections to identify areas where the species were likely to expand, contract, or remain unaffected. Cheatgrass distribution, for example, is primarily limited by spring and summer precipitation, annual precipitation, and maximum winter temperature. Combining this information with regional climate projections indicates an increasing risk of invasion in Idaho, Montana, and Wyoming, but decreasing risk in southern Nevada and Utah. Much of the area currently occupied by cheatgrass in the central Great Basin is likely to become increasingly unsuitable for it over the next century, which could open up opportunities to rehabilitate currently invaded areas by replanting species that grew there previously or species expected to move naturally into the area as a result of climate-driven range changes.

**BOX 12.2 WHEN IS A NONNATIVE NOT A NONNATIVE?**

Around the globe, species' ranges are expanding or shifting in response to climate change. As species naturally establish themselves in new areas, should they be eradicated as unwanted invaders or welcomed as climate refugees? While range shifts are generally beneficial for the species doing the shifting—tracking suitable conditions is better than staying put and dying out—the arrival of a new species may be bad news for existing species or communities if newcomers prey on, infect, or outcompete old-timers for resources.

A related debate is whether assisted migration (managed relocation) of species is a reasonable conservation tool or an unwarranted disruption of native ecosystems (chapter 9). Again, the response to this debate depends on whose perspective you take: the species on the move, or the species being moved in on.

High-risk areas can also be identified by investigating regional abiotic factors that correlate with invasion likelihood. For instance, alien species richness in New Zealand forests is highest in small, isolated forest fragments in warmer, drier climates (Ohlemüller et al. 2006), suggesting that forest fragments will face increasing invasion risk as warming progresses. On an even broader scale, rapidly changing regions such as the Arctic may be more vulnerable to invasion because the high rate of change puts natives at a disadvantage (Byers 2002).

*Species Interactions*

Species' success is strongly influenced by interactions with other species. If warming is better (or less worse) for an invader, its prey, or its mutualists than for its competitors, parasites, or predators, warming will likely enhance its overall success. Nonnative mussels in San Francisco Bay appear more tolerant of high temperatures during exposure at low tide than native mussels, for example, so would be favored by increasingly warm conditions. Nonnative tunicates in New England become increasingly dominant over natives as water temperature rises, most likely because the invaders appear earlier in the spring and recruit more heavily following warm winters, while natives have lower recruitment and no change in appearance time (Stachowicz et al. 2002).

Native species stressed by climate change could be weaker competitors for space or other resources, or be more vulnerable to pests and pathogens. Rapid change will also favor invasive species with characteristics allowing them to track or adapt to rapid environmental change, such as shorter generation times, rapid dispersal, or high fecundity (Theoharides and Dukes 2007). Nonnative springtails on sub-Antarctic Marion Island, for example, are more thermally tolerant and have faster generation times than native springtails. Differences in evolutionary rates may be particularly important for host-pathogen systems. Disease-causing organisms typically have rapid evolutionary rates, as do many insect species that are common vectors for disease. To the extent that they are better able to respond to warming than their hosts, there may be at least a temporary shift in favor of disease-causing organisms.

### *Invasive Species and Climate Vulnerability*

Many nonnative species significantly alter the physical or chemical properties of the areas they invade, in some cases amplifying and in others diminishing the changes brought by climate change. A number of plant species, for instance, are likely to increase drought stress. Some invasive grasses appear to decrease soil moisture more rapidly than native grasses, and saltcedar, a highly invasive woody shrub, consumes up to 200 gallons of water a day (Di Tomaso 1998), lowering groundwater levels and drying up springs and marshes. In contrast, the presence of domestic cattle (a New World nonnative) may actually decrease the vulnerability of vernal pool ecosystems in parts of California to climate change–related drought (Pyke and Marty 2005). Pools where cattle were excluded dried out on average fifty days earlier than pools where grazing was allowed, presumably because grazing kept plant growth (and thus water use) under control.

Nutria, a species of South American rodent, has become common in coastal North America and is likely increasing the vulnerability of coastal areas to sea-level rise in a number of ways. First, these animals facilitate the conversion of marshes to open water by destroying vegetation and root mats, decreasing natural defenses against erosion and flooding due to sea-level rise. Nutria also create extensive depressions and swim canals that allow saltwater to move more quickly into previously fresh areas. Tens of thousands of acres of coastal wetland along the Gulf of Mexico alone have been degraded by nutria. Eradication of nutria, as has been accomplished in California, helps to prevent the loss of natural coastal defenses against sea-level rise and storm surges and could thus be an important element of decreasing coastal vulnerability to climate change in coastal regions.

Wildfires are expected to increase as a result of climate change in western North America. Wildfire risk is also increasing in this region because of an extensive, ongoing invasion by nonnative cheatgrass. As cheatgrass moves in, it converts woody shrublands into annual grasslands that are more susceptible to fires. The combined climatic and biotic effects on wildfire frequency and intensity could push western ecosystems permanently into a new state.

Finally, nonnative species may increase the vulnerability of natives to climate change by restricting the habitats available to them. Nonnative trout in the Sierra Nevada mountains in the western United States prey heavily on the endangered mountain yellow-legged frog, excluding these amphibians from many of the larger, deeper lakes that provide climatic refugia during warmer, drier years (Lacan et al. 2008; fig. 12.3).

### *Climate Change and Purposeful Introductions*

As with unintentionally introduced species, climate change will likely influence the success of nonnative species introduced for biocontrol, commercial, or other purposes.

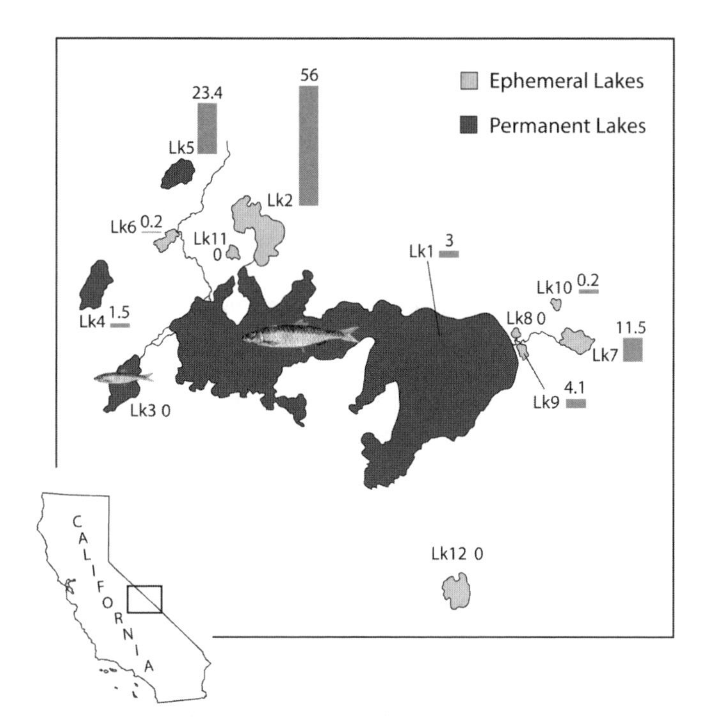


FIGURE 12.3 Percentage (as shown by bars) of total Sierra Nevada yellow-legged frog egg masses found in lakes in Upper Dusy Basin in the Sequoia and Kings Canyon National Parks. Lakes 1 and 3 have large populations of introduced trout. The majority of egg masses occur in an ephemeral lake (Lake 2), and trout are effectively excluding frogs from using the lake least vulnerable to climate change (Lake 1). After Matthews et al. 2009.

When Pacific oysters were introduced to the western shores of Canada and the United States, for instance, it was expected that they would not become invasive because cool water temperatures would prevent them from reproducing in the wild. Periods of generally warmer waters over the last half century have led to the oysters' rapid spread, displacing native rockweed communities in Washington State and mudflat communities on Vancouver Island (D. Padilla and C. Mills pers. comm.). A similar story has played out with nonnative oysters in the northern Wadden Sea. Although an oyster farm was established in 1986 and nonnative oysters sporadically expanded throughout the region in the following years, dense oyster beds became established only after two particularly warm summers in the early 2000s (Diederich et al. 2005).

Climate change may also influence the effectiveness of biological control efforts against invasive species, such as the use of predators and parasites against the recently introduced vine mealybug in California, a major pest of commercial vineyards. Projected warming throughout California is generally favorable for three species of insect introduced for biological control of mealybugs, but is also favorable for the mealybugs themselves. Interactions among a range of factors will likely lead to a decrease in the effectiveness of biological control not just for vine mealybugs, but for other pest species as well (Gutierrez et al. 2008). Incorporating climate change considerations into the selection of biological control agents may increase the effectiveness and limit the risks of biological control.

As global momentum builds to limit climate change and its effects, people may be tempted to jump into projects that involve nonnative species without fully considering

**BOX 12.3 POLICY CONSIDERATIONS**

The range of interactions between climate change and invasive species highlights the importance of ensuring some level of policy coordination around these issues. Pyke et al. (2008) suggest three possible policy filters: (1) climate change mitigation activities should not exacerbate invasive species problems (e.g., nonnative invasive species should not be planted for use as biofuel); (2) invasive species management should take climate change into consideration (e.g., invasive species that reduce resilience to climate change should be prioritized for control); and (3) efforts to reduce vulnerability to climate change should not make the invasive species problem worse (e.g., projects to reduce coastal vulnerability to erosion should not use nonnative species).

the potential for unintended consequences. For instance, the burgeoning biofuel movement may enhance invasions by increasing plantings of nonnative crops. At least two nonnative plants already considered major threats to native wetland and riparian habitats, giant reed and reed canary grass, are under consideration as potential feedstock in the United States, and the Department of Energy's 2009 *Multiyear Biomass Program Plan* makes no mention of the possible risks of using invasive or nonnative species for feedstock. This lack of awareness could lead to federal funds being spent simultaneously to plant species for biofuel and to eradicate those same species as noxious weeds.

Similarly, increased coastal erosion combined with growing awareness of the negative effects of shoreline hardening (e.g., bulkheads or riprap) is leading to increased calls for biological shoreline stabilization. Past experience suggests that extreme caution should be used in considering the use of nonnative species to provide this and other ecosystem services. For instance, ice plant, European beachgrass, and smooth cordgrass were introduced along the west coast of North America to stabilize dunes and minimize erosion. These plants have become widely established, threatening a variety of native species, and are now the focus of extensive and expensive eradication campaigns. In many cases, native plants or animals can provide the same stabilizing function.

## Diseases and Pests

Climate change is likely to alter pest and disease dynamics, regardless of whether the organisms in question are native or nonnative. A well-known example is that of the native mountain bark beetle in North America. Normally limited by winter cold snaps, beetle populations have expanded to higher elevations and latitudes as winters warm. Due to the high elevation at which they live, whitebark pines had previously escaped extensive outbreaks and have no natural defenses against the beetle. As warming increases the beetle's elevational range, whitebark pines have suffered significantly, and Canada's northern jack pine forests may also be at risk. Outbreaks in the beetles'

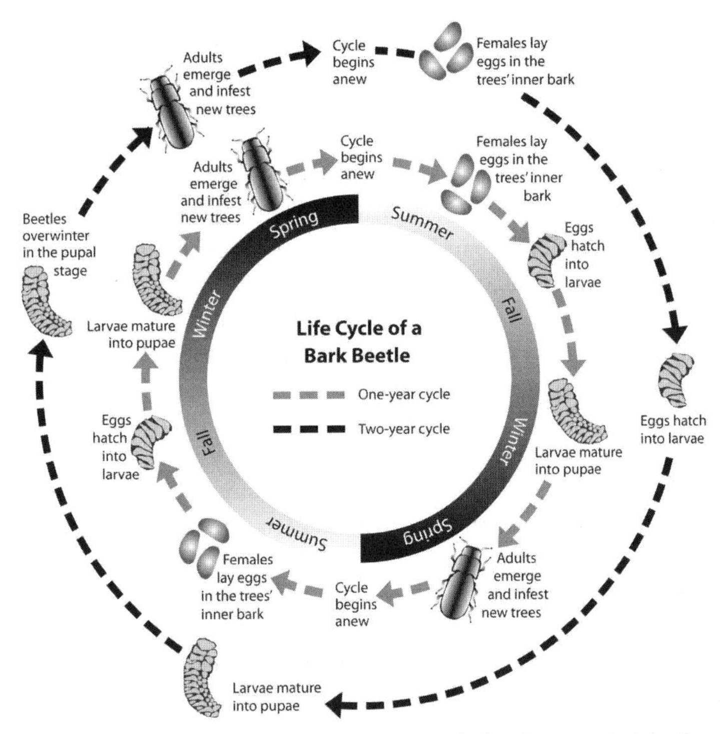


FIGURE 12.4 Pine bark beetle life cycle. The two-year cycle has been typical in the past, but warmer weather is making the one-year cycle more common.

historic range have become more frequent and severe as warmer summers allow them to complete their life cycle in one year rather than two (fig. 12.4). Although management options for bark beetles are limited, projected changes in range and outbreak severity can be used to inform management decisions and to prepare human communities in at-risk areas for potentially large economic and ecological changes from dying forests and increasing numbers of wildfires. A similar story is playing out with the oyster diseases dermo and MSX along the east coast of the United States (Ford and Smolowitz 2007). Although these diseases were once rarely found north of Chesapeake Bay, warming ocean temperatures have allowed them to expand north to Maine and Nova Scotia.

The discovery of a previously undescribed lungworm in Canadian muskoxen in the late 1980s appears to reflect the combined effects of climate change and species recovery. Although this lungworm was unknown before the 1980s, it is unlikely that the parasite itself is new to muskoxen. Most likely it was present in relict populations that survived periods of overhunting, and as these populations expanded when hunting was stopped, the parasite expanded with them (Kutz et al. 2004). Adult lungworms up to 65 centimeters long live in cysts in the lungs of their primary host, muskoxen. Larvae crawl up the trachea, are swallowed, pass out in feces, and move into their intermediate host, slugs. Larvae go through several stages in the intermediate host before they are capable of reinfecting muskoxen. Typically, this is a two-year process, with larvae overwintering in their slug hosts. Longer, warmer summers speed up development rates and give the larvae a longer season in which to reach the transmission phase, so many larvae now become available for transmission in their first year. Warmer winters mean lower mortality of slugs and larvae, increasing the number that can infect muskoxen the next year. In Norway, outbreaks of a related parasite are linked with warmer-than-average years, and there is anecdotal evidence linking increased parasitism with muskoxen population declines in the Canadian Arctic.

The spread and impact of pests and diseases will also be influenced by effects of climate change on potential hosts, including changes in immune status, body condition, toxicant exposure, and population density. For instance, if a decrease in available feeding area creates increased crowding in the remaining areas, density-dependent transmission of parasites or diseases would increase. And just as people are more susceptible to disease when stressed, plants and animals that are stressed by changing climatic conditions are also likely to be more susceptible to pests and diseases.

The need to respond to increased pest and disease outbreaks will lead to its own set of challenges. There may be calls for widespread use of pesticides or other drastic control efforts if economic stakes are high, as is the case with forest resources. In the case of critically endangered species at risk of exposure to novel pathogens or parasites, conservation practitioners may decide that it is worth the time and effort to carry out extensive precautionary vaccination, or to monitor wild populations and provide captive or in vivo care to individuals showing signs of illness. Before leaping into action, managers would do well to consider the longer-term feasibility, cost, and effectiveness of such efforts, as well as potential negative effects. Climate change is not a short-term phenomenon, and while short-term coping strategies may be appropriate in some cases, they are rarely the best long-term response strategy.

## Final Thoughts

Interactions between climate change and invasive, pathogenic, or parasitic organisms offer a wealth of challenges, both practical and philosophical. At the very least, an awareness of these challenges allows us an opportunity to develop strategies proactively rather than reactively, and to establish guidelines and communication pathways that may limit our vulnerability as things heat up. There may even be situations in which climate change facilitates the control of invasive species (as with Argentine ants in California), or opportunities where targeting particular nonnative species can greatly decrease ecosystem vulnerability to climate change (as with nutria). Knowledge is power, if used wisely.

# PART III

## *Rethinking Governance, Policy, and Regulation*

Some of the actions we need to take to adapt our work to climate change may in fact go beyond the scope of how we normally envision our resource management and conservation work. Climate change impacts do not stop at park boundaries, and effective conservation and management may need to think beyond boundaries as well. To some degree this has always been true: climate change just makes it more pressing.

One of our most effective cross-boundary conservation and management tools has been strong regulations, particularly of pollutants (chapter 14) and of the rates at which fish, forests, and other resources are harvested (chapter 13). What happens when pollutants become more toxic because of changing environmental conditions, or when a fishery moves out of its historically regulated region into a region where it is not currently managed? The world is changing, and how we think about and use regulatory tools will need to change as well. Just as business as usual is not a good idea for greenhouse gas emissions, it is also not a good idea for conservation and management.

To be most effective, existing tools for conservation and resource management need to be put in context. While tensions between development and conservation are not new (chapter 15), the simultaneous adjustment of human and ecological systems to a changing climate offers a new set of pitfalls and opportunities. The governance context in which decisions are made (chapter 16) strongly influences the ability of planners and managers to operate in a climate-savvy manner. Enabling conditions for the effective incorporation of climate change into resource management and conservation include an adaptive governance framework that supports the development, evolution, and use of evidence-based management over the long term. Success is unlikely to come from one-time decisions made on today's best available data; rather, it will come as we try out different options, see if they work, and adjust accordingly.

As you reach the end of the book, we encourage you to remember to use the best tool of all—your brain. As is often said, you can't solve a problem with the same thinking that caused the problem. Relying on the same methods we have always used despite the change in conditions is not likely to yield success. Conservation and management require creative new thinking as well as data. Creativity is crucial to effectively adapt our work to climate change.

# Chapter 13

# *Regulating Harvest in a Changing World*

> If you wait fifty years with your worms and your wishes,
> You'll grow a long beard long before you catch any fishes.
> —*Theodor Geisel (Dr. Seuss)*

Images of clear-cut forests often serve as shorthand for environmental degradation, but such unsustainable overharvest manifests in many other forms including overfishing, enormous bycatch, and overallocation of freshwater resources. Although unsustainable use or harvest initially appears as a local disturbance, its effects ripple far from the center of destruction, cascading through food-webs and ecosystems. Often, these consequences compound or are compounded by the adverse effects of climate change. Levels of resource use that are sustainable now may not continue to be sustainable, for example, and overharvest can worsen the effects of climate change or even increase the rate of change itself.

Reconsidering when, where, and how we extract natural resources may help us develop management practices and policies that reduce the vulnerability of the resources, resource users, and related ecosystems to climate change. Reducing harvest levels or shifting harvest location and timing in response to climate change effects can increase population and community resilience, supporting connectivity and maintaining populations large and genetically diverse enough to buffer against unexpected effects of climate change.

## Redefining Sustainable Use

Overharvest is an old problem with well-known effects such as loss of biodiversity and evolutionary potential, damage to food webs and physical habitat, and a host of other potentially negative consequences. Some marine species have been fished to functional or economic extinction, including classic examples from whaling and sealing operations as well as more contemporary examples such as tuna, swordfish, sharks, and cod (e.g., Jackson et al. 2001; fig. 13.1). In extreme cases, overharvest can lead to complete species extinction due to direct harvest (e.g., Caribbean monk seal) or a combination of direct harvest and deforestation or other habitat loss (e.g., passenger pigeon).

All of these problems magnify the consequences of climate change. Most basically, reduced numbers of individuals or species decrease the ability of a population, community, or food web to successfully respond to disturbance, including climate-driven effects such as shifts in food supply, temperature, or water chemistry. Decreasing numbers of individuals or subpopulations reduces connectivity and increases the vulnerability of populations or species to extinction. Also, smaller populations typically have less genetic diversity, reducing their evolutionary options. While today's climate change is happening quite rapidly, evolutionary adaptation will still play an important role in helping some populations or species survive or even thrive in the new climate regime. Such evolutionary adaptation depends on the presence of individuals with the right genetic characteristics, however, and by reducing the pool from which these lucky winners may emerge, some opportunities are lost.

As mentioned in chapter 5, climate change could alter the population dynamics of many species, affecting key variables such as number of offspring, food availability, or predation. Failure to account for these changes could lead to unintentional

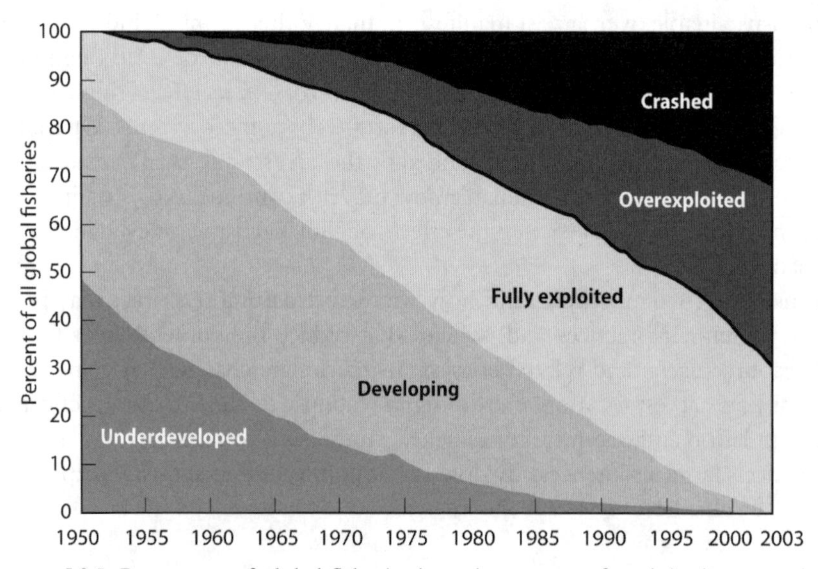


FIGURE 13.1 Percentage of global fisheries in various states of exploitation over time. After U.N. Environmental Program 2007, figure 4.13.

overharvest, unnecessarily restrictive harvest limits, or failure to create balanced harvest of species in mixed-species harvest regimes where different species respond differently to climate change. Frameworks for incorporating climate change into harvest regimes are being developed (e.g., Hollowed et al. 2009; A'mar et al. 2009), but much more refining and field-testing is needed.

### BOX 13.1 CHANGING FISHERIES MANAGEMENT IN THE BERING SEA

The Bering Sea is highly productive, providing nearly half of the annual seafood catch of the United States. In addition to its rich fisheries, many marine mammal species and millions of migratory birds feed in the Bering Sea for at least part of every year. Aboriginal groups and rural households rely on these natural resources for subsistence. Dramatic warming has caused equally dramatic declines in seasonal ice coverage, with ecosystem-wide consequences. Managers and legislators can do little to directly compensate for sea ice loss, but there are actions that can slow ice loss (mitigation of greenhouse gas emissions), as well as actions to increase ecosystem resilience (e.g., adjusting fishing to levels that can be supported under the new climate regime and adjusting fisheries management to the northward shift of fish stocks).

By unanimous approval in June 2007, the North Pacific Fisheries Management Council (NPFMC) designated a northern boundary for bottom trawling based on essential fish habitat for the Bering Sea, and in January 2009 it approved a fisheries management plan that prohibits commercial fisheries in the U.S. Arctic "until adequate scientific information on fish stocks and how commercial fisheries might affect the Arctic environment are available" (NPFMC 2009). This plan grew out of the council's recognition of "heightened national and international interest in the Arctic and potential changes in this region that might arise due to climate warming."

Two programs in particular are helping to build climate change into Bering Sea fisheries management. The North Pacific Climate Regimes and Ecosystem Productivity (NPCREP) study uses monitoring, modeling, and experiments to investigate how climate variability and change affect the physical and biological controls on ecosystems in this region. This information is used to develop indices and assessment tools the NPFMC can use in determining each year's total allowable catch, as well as fish recruitment predictions that include the effect of climate change. The NPCREP program also provides online access to environmental and ecosystem data for the Bering Sea that allow the NPFMC to track trends that feed into management recommendations. A related program, the Bering Sea and Aleutian Islands Integrated Ecosystem Research Program (BSIERP), also generates and provides data that will help in the management of fisheries, marine mammals, and seabirds. The BSIERP project also works to document, characterize, and quantify local subsistence and cultural use, as well as indigenous understanding of the Bering Sea ecosystem, and to integrate this knowledge into ecosystem models.

The concept of sustainable use applies to nonliving as well as living resources. Human demand for water has increased over the years in response to population growth, changes in agricultural practices, and other forces, and water is now being removed from many rivers, lakes, and aquifers faster than they are being refilled. The Ogallala Aquifer, which provides 30 percent of all groundwater irrigation in the United States, has been severely depleted: water levels in parts of the aquifer have dropped by more than 150 feet in the last half century. Climate change simulations for the region vary, but all predict further reductions in aquifer recharge (Rosenberg et al. 1999). Thus current rates of water use will deplete the aquifer even faster than before. As with the Colorado River (see chapter 3), we must consider the changing availability of resources—especially water—as the climate changes, and adjust their use and extraction accordingly if we want continued access to them in the future. Such adjustments typically include a combination of decreased overall extraction and increased efficiency of use.

## Droughts, Floods, and Pestilence

Overharvest itself can cause climatic change. On a local scale, clear-cutting forests causes warming and drying due to loss of shade and altered hydrological cycling. This problem is particularly pronounced in the tropics, and farmers in Africa and elsewhere have realized that by allowing some trees to grow in their fields they can decrease drought and increase yield. Clear-cutting can also affect regional climate: the warming and drying that have caused extinctions in Costa Rica's cloud forests result from a combination of global climate change and lowland deforestation. Lowland forests supplied significant moisture to the air that flows up and over the mountains, feeding the cloud cover that supported a rich forest ecosystem, but much of that forest has been converted to agriculture.

Deforestation even affects the global climate. Decreased forest cover by itself means decreased carbon uptake and storage, increasing the rate at which carbon dioxide builds up in the atmosphere; when fires are used to clear forests, the effects can be even stronger. During the 1997–1998 El Niño, which created extremely dry conditions in some areas, forest-clearing fires in Indonesia ran out of control and the combustion of both forests and rich peat soils emitted greenhouse gases equivalent to 13 to 40 percent of fossil fuel combustion that year (Page et al. 2002). These emissions contributed to global climate change, while the smoke changed weather patterns for thousands of miles and the loss of forest changed regional climate patterns.

While forests remove greenhouse gases from the atmosphere, it is possible for clear-cutting to have a cooling influence if the new vegetation cover absorbs less heat than the forests it replaces. Models indicate that global replacement of grasslands with trees could warm the planet by up to 1.3°C, while replacing forests with grassland results in cooling of roughly 0.4°C (Gibbard et al. 2005). Clearly, getting rid of all forests is not a good conservation plan, as the innumerable negative effects of such a

**BOX 13.2 THE BIGGEST OVERHARVEST ISSUE OF ALL**

**While overharvesting fish and trees can compound the adverse affects of climate change, it is overuse of fossil and forest fuels that is at the root of the problem. Wood can be a renewable resource given proper forest management, but fossil fuels are not renewable on human timescales. It is useful to consider societal attitudes about harvest and use of all natural resources when constructing solutions to both the causes and effects of climate change. At some point, our profligate use of these precious resources must be resolved if we are to develop sustainable solutions to the many problems caused by overuse.**

choice far outweigh any benefits for mediating global temperature. To paraphrase Ken Caldeira, we should focus on stopping climate change to save the forests, not saving forests to stop climate change.

This highlights the danger of assessing our choices and actions through a single lens. Here, the benefit of reducing global temperatures by replacing all forests with grassland is more than balanced by the loss of habitat biodiversity and the ecosystem services provided by forests. In a similar fashion, many are concerned that the use of biofuels as a strategy to reduce climate change could lead to biodiversity loss, widespread introduction of nonnative invasive species, and higher food prices. We must continue to weigh the costs and benefits of each measure in a holistic fashion, and make sure decisions result in an overall net gain of sustainability for the planet and ourselves.

Loss of forest cover can contribute to flooding as well as drought. During rainstorms, intact forests slow the rate and volume of water runoff, meaning more moisture stays in the forests for gradual release later and less floods straight into streams, lakes, and rivers. Over the longer term, forest loss increases sedimentation of rivers and streams, shrinking the volume of water they can hold before overflowing. Thus in areas where climate change is likely to cause an increase in heavy rains, adjusting harvest levels and techniques to account for local flood and erosion risk can help to reduce vulnerability. Similarly, reducing overharvest of mangroves in coastal areas can decrease erosion and increase sediment retention, reducing the rate at which shoreline is lost to rising seas.

Reducing harvest is not the only path to adaptation: strategic shifts in the timing, location, or methods of harvest, or even increasing harvest in some situations, may also help. One example is the pine beetle infestations of forests in North America stretching from Colorado to Alaska. The government of British Columbia has proposed a strategy for both economic and ecological protection by *increasing* harvest (Nelson 2007). The first phase was to shift from harvesting healthy trees to harvesting infested trees to limit the spread of infestation. The second phase was to harvest dead and weakened trees. This salvage phase has an economic interest—harvest of timber for sale—as well as a management interest—reducing fire risk from massive stands of dead trees. Some retrospective discussion has centered on whether more dramatic harvest early in the outbreak might have limited the area affected, while others have proposed that managing for maximum yield suppressed natural fire regimes and made the forests more vulnerable. Even with natural fire regimes, however, it may be that temperatures no longer get cold enough to suppress pine beetle infestations in some parts of their range. Discussions of triage and engineered approaches to climate change adaptation will require continued exploration of the role of harvest and other proscriptive actions, although many find them counter to traditional conservation principles.

## The Web of Life

As is clear from the deforestation examples above, harvest levels and techniques can have effects well beyond the target place or species. Gill nets targeting a range of fish species also kill hundreds of marine mammals and turtles each year. For every pound of shrimp that shrimp trawlers keep, they typically bring up 8 to 10 pounds of other species that are simply thrown overboard dead or dying (Davies et al. 2009). The loss of wolves throughout much of their original range in the United States allowed deer to flourish, reshaping native forests in ways that may increase their vulnerability to climate change. Determining harvest levels or resource allocation must be done with an eye toward these indirect effects and their influence on system-wide climate vulnerability. In some cases this may lead to a need for increased harvest levels (e.g., deer), in others to decreased levels.

The problem is deeper than just reduced population sizes or species loss. We are "fishing down the food web," harvesting species from higher trophic levels to the point of economic extinction, then moving down to the next trophic level (Pauly et al. 1998). Climate change may make it more difficult for these overfished systems to return to their previous state. For instance, the tenfold increase in Bering Sea jellyfish in the 1990s may be partly linked to climate change, and these jellies will reduce the food available for larvae of many commercially harvested species in the area as well as consuming the larvae themselves. An explosion in jellyfish populations in the Black and Asov Seas in the 1980s, while not directly linked to climate change, virtually wiped out once-productive fisheries there.

On coral reefs around the world, the loss of herbivores due to overharvest or disease is decreasing reef resilience to climate change. In the absence of grazers, mass coral-bleaching events are followed by an explosive growth of seaweed that makes it difficult for the reefs to recover (Hughes et al. 2007). Supporting healthy herbivore populations is thus an essential element of avoiding a shift from coral- to algal-dominated systems under climate change.

This interaction of climate change with overharvest and other stressors is also playing out in the Chesapeake Bay. The initial collapse of oyster populations resulted primarily from overexploitation, but poor water quality, climate change, new diseases, and

### BOX 13.3 DON'T PICK JUST ONE

**Although this book discusses categories of adaptation options in separate chapters, an adaptation strategy should encompass multiple options. For example, the salmon discussed in this chapter need streams with cool water and good gravel habitat, plenty of food in the ocean, and large enough populations to insure against occasional disaster. Protecting them will require not just limiting salmon harvests but also potentially removing dams, decreasing the human demand for water from salmon-bearing rivers, maintaining and restoring riparian vegetation, protecting water quality, and a host of other approaches. It is important to consider how each of these factors will be affected by climate change and to adapt our strategies accordingly. Adjusting harvest levels, timing, and techniques is one approach for creating more robust systems.**

interactions among these stressors have prevented its recovery. The loss of the oysters' immense water filtration capacity (at its peak the Chesapeake oyster population is said to have filtered the entire bay in a single day) combined with increasing nutrient pollution and warmer water is causing massive phytoplankton blooms, leading to a growing hypoxic "dead zone." Warmer water due to climatic changes has also allowed southern oyster parasites to expand their range northward into the bay. Finally, efforts to restore sea grass and invertebrates are hampered because the bay may no longer be climatologically suitable for species that once called it home. It may be that overharvest, pollution, and climate change have pushed Chesapeake Bay into a new state from which it will be difficult to recover.

Climate-savvy harvest management may also be critical for protection of places and resources that seem unrelated to the stock in question. For example, depletion of salmon populations from the ocean affects not only marine food webs, but also the streams where salmon spawn and the forests that line the riverbanks. Salmon bring nutrients from the oceans back to the streams where they spawn, die, and decay, releasing nutrients directly into the water column or indirectly to the surrounding terrestrial ecosystem through the work of scavengers. Gresh and coauthors (2000) estimate that nutrient input from salmon in the United States' Pacific Northwest is just 7 percent of historical levels. This has caused shifts in production and composition of stream, lake, and riparian communities (e.g., Naiman et al. 2002). Whereas the Chesapeake Bay suffers from too much nutrient input, the problems in these streams stem from too little. Just as with the Chesapeake Bay, however, warming conditions and altered water flow due to climate change may further compound the community shifts. This is particularly true for the Pacific Northwest, given the likely negative effects of climate change on salmon populations in that region. Reducing salmon harvest, removing dams, and other measures to increase salmon success may be some of the best bets for not only increasing salmon resilience, but also affording their freshwater habitat some buffer to the effects of climate change.

## Shifting Time and Space

Many harvested species already are or will be exhibiting range shifts. For example, North Pacific pollock distribution shifted significantly northward between 1999 and 2007. In the Bering Sea and Arctic Ocean, some fisheries management councils are beginning to grapple with what this will mean for the future of fishing (see box 13.1). During past ice ages many tree species shifted their ranges across entire continents, and there is some evidence that tree populations at the warmer ends of their ranges are already suffering from warming trends. Maximum sustainable yields will change for different regions as populations move into or out of traditional management areas. Growth rates, a key component of many fisheries harvest models, will also change with climate change. For example, models suggest that the yield of walleye in Ontario, Canada, will increase in the north and decrease in the south in a warmer world. Limits on fisheries will need to change to protect species or ensure sustainable yields as locations and population dynamics shift in response to changing climatic conditions.

Climate change is creating temporal as well as spatial change, such as changes in when seasonal events happen and increased variability in populations or resource availability over time. In many cases, harvest regulation and management is already designed to cope with variability. The Pacific Coastal Pelagic Fisheries Management Plan already adjusts harvest for periods of high or low productivity, such as for sardine stocks in relation to Pacific Decadal Oscillation or El Niño–Southern Oscillation cycles (fig. 13.2). Where seasonality and variability are not already taken into consideration, managers should at least assess the importance of doing so.

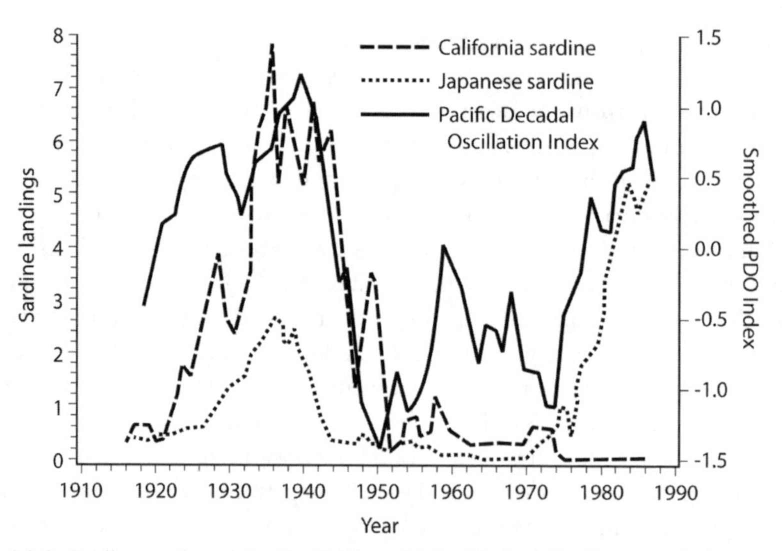


FIGURE 13.2 Sardine catch and the Pacific Decadal Oscillation. Sardine populations tend to be high when the Pacific Decadal Oscillation index is greater than 0 (warm phase) and low when it is less than 0 (cool phase). The correlation between population size and climate regime allows temperature to be factored into harvest rules. Sardine landings after FAO 2005.

## Final Thoughts

Definitive coverage of issues relating to climate change and harvest are beyond the scope of this book. Rather, we hope the range of examples provides a sense of the multifaceted nature of the problem, and catalyzes thinking about equally multifaceted solutions.

Preventing the damage of overharvest has been a pressing issue for generations, and climate change only promises to compound the challenge. But there are opportunities in how we address resource extraction to create more climate-robust management schemes. Centuries of exploitation have resulted in the "shifting baseline" phenomenon whereby each generation accepts a diminished level of biodiversity or abundance as the new normal. Climate change threatens to be the ultimate shifting baseline. The challenge is to limit that shift by including climate-savvy harvest management in our strategies. Whether it is trees, fish, or some other harvested resource, planning ahead for changes will likely yield better results than waiting for dramatic changes to occur and responding in a reactionary fashion.

# Chapter 14

## *Regulating Pollutants in a Changing World*

The truth is rarely pure and never simple.
—*Oscar Wilde*

Societies have made great strides in reducing the damage of environmental pollutants by enacting regulations, implementing testing and monitoring criteria, and developing treatment strategies to reduce toxicity and damage. Although new challenges do arise (e.g., low-dose toxicity issues with some plasticizers), efforts to date have led to higher air, water, and soil quality in many regions. Climate change, however, points to a glaring limitation of our current regulatory system. The testing procedures on which current regulatory limits are based do not generally reflect real-world exposure conditions, and certainly do not reflect a world in which temperature, salinity, and a host of other factors are changing as a result of climate change. Although regulatory limits for some pollutants specifically address season or local water chemistry, this is not the norm.

Through its alteration of environmental conditions and our responses to those changes, climate change is affecting the exposure, uptake, and toxicity of environmental pollutants for humans and ecosystems alike (e.g., Noyes et al. 2009). Pollutants are also affecting the vulnerability of species or ecosystems to various elements of climate change. Adapting pollutant use and regulation to climate change is thus a double-sided coin. We must consider how to adapt pollutant use and regulation to remain effective under changing climatic conditions, and how to decrease human and ecosystem vulnerability to climate change by adjusting our use and regulation of toxic compounds.

**BOX 14.1 CONTAMINANTS IN THE REAL WORLD**

Decades of toxicity testing have been used to set regulatory limits on a broad array of environmental contaminants. Strict protocols were developed to standardize the tests and allow for comparisons among different pollutants and their effects. Unfortunately, these protocols miss key elements of the real world such as day-to-night temperature fluctuation or the adaptation of different populations to different conditions. Climate change adds still more complications. Not only will environmental conditions be changing, but species may also be adapting evolutionarily, behaviorally, or physiologically to the new conditions. Current and future variability in temperature, salinity, and other environmental variables may mean that some regulatory limits are or will become insufficient.

Furthermore, testing protocols typically fail to address the issue of spectrum or intensity of lighting, despite the fact that these variables can affect both chemical toxicity (some compounds change to more toxic forms when exposed to full-spectrum sunlight) and organisms' sensitivity to it. Most toxicity assessments are conducted under ambient laboratory lighting with little if any ultraviolet (UV) irradiance. Here, again, climate change highlights the importance of this omission. In some areas, climate change may allow UV to penetrate more deeply into the water, for instance due to a reduction of colored dissolved organic matter by acidification. In others, it may decrease total light levels, including UV, due to increased terrestrial runoff.

Finally, it is important to note that much toxicity testing is done on temperate species under roughly temperate latitude conditions. Tropical, polar, and desert regions are not within the range of the standard exposure parameters, and few species from these regions have been used for testing.

## Changes in Exposure and Availability

An important component of vulnerability to pollutants is the potential for exposure and the ease with which any given pollutant is taken up by organisms once they are exposed to it (known as the *availability* of the pollutant). Exposure is typically linked to the amount of pollutant present in the environment and how rapidly it breaks down, while availability is related to the form a compound is in or the medium (water, soil, air) in which it is found. Availability can also be affected by other environmental conditions, such as pH or temperature. Climate change, through its direct effects on the environment as well as human responses to it, will affect both the concentration and the availability of a range of environmental pollutants.

### *Changing Sources*

Which pollutants are used where will change as agricultural opportunities, weeds, pests, and vector-borne diseases shift geographically in response to climate change.

Areas where cold winters or short summers have previously provided sufficient pest and disease control may turn to chemical control as warming proceeds or as new pests—including both insects and microbes—arrive in the area thanks to more favorable environmental conditions.

The expanding ranges of agricultural and disease-carrying pests are likely to lead to increased use of chemicals, particularly from widespread application of insecticides near insect breeding areas and threatened human communities. For example, New York City has carried out citywide aerial pesticide spraying in response to the West Nile virus, and California has suggested statewide aerial spraying to combat the nonnative light brown apple moth. Increased individual use of insecticides and medications, exacerbated by aging human populations and growing global use of pharmaceuticals, may also contribute, as many of these compounds end up in receiving waters from municipal wastewater treatment discharge. The need to control the spread of disease must be weighed against the environmental effects of these compounds. In the case of New York City, for instance, it is unclear that the use of insecticide had any significant effect on the incidence of West Nile virus (Thier 2001).

Finally, ecosystems, species, or individuals that move into new areas may be exposed to new contaminants or new sources of contaminants. The highest risks in this regard involve polluted areas near the coast that may be periodically or permanently flooded as sea level rises, releasing chemicals into the marine environment. Increases in flooding or storm strength could increase the risks of pollutant spills across biomes.

### *Changing Use*

Many pests and diseases respond favorably to warmer conditions. For instance, insect populations are often held in check in temperate areas by die-off during winter cold snaps or by the number of generations they can squeeze into warm summers. As spring comes earlier and fall later, more generations can be produced each year, and warmer winters mean more individuals survive until spring. Thus there may be shifts in application frequency and concentration of pesticides, herbicides, fertilizers, or other chemicals. For instance, fungicide use may increase in areas that experience longer, wetter periods, and pesticide use may increase in areas where warmer conditions cause insect populations to increase. All of these chemicals will run off into the surrounding environment, affecting the surrounding landscape and associated receiving waters.

Chemicals may also be used more in forestry in response to increasing severity of frequency of fire, pest, and disease outbreaks. There has been some discussion of trying to control the increasing incidence of pine beetle in North America with more liberal preventive application of pesticides, although the amount of pesticide needed to address the affected area and the low likelihood of long-term success make this an unappealing option. The predicted increase of wildfires as a result of climate change in many regions will likely increase the use of flame retardants, some of which are highly toxic. While this toxicity is widely recognized and two classes of these compounds have already been banned in Europe, Canada, and the United States, others are still in use

(Ross et al. 2009b). It is projected that these compounds will be the most common contaminant in the tissue of fish and marine mammals, surpassing PCBs, within the decade even without increased use due to climate change.

Although higher levels of atmospheric carbon dioxide may have some initial benefits for agriculture by essentially fertilizing crops, most systems become nitrogen-limited as carbon dioxide enrichment occurs. This may lead to increased use of nitrogen fertilizer to maximize yields, contributing to inland waterway degradation, toxic algal blooms, and coastal dead zones—all of which are only made worse under elevated temperatures.

### *Changing Transport*

Contaminants are carried by water, air, and soil movement, all of which will be affected by climate change and our responses to it. The nature of these effects may be complex. More severe flooding and runoff could increase the amount of pollutants washed into aquatic systems (e.g., fig. 14.1), but it may also dilute those same pollutants and reduce their impact. During prolonged droughts, there may be more airborne drifting of contaminated soils or a greater buildup of pollutants on land and in the soil. The first rains following prolonged drought may deliver particularly concentrated pollutants to aquatic systems. Further confounding issues around water-mediated toxicity will be changes in how water is used by natural and human systems under climate stress, for instance irrigation rates, metabolic shifts in aquatic species, changes in peak runoff periods, or changes in animal reproductive timing. Finally, warmer conditions may enhance the volatilization of some compounds.

Climate change may also bring some exposure surprises. Although DDT and polychlorinated biphenyls (PCBs) have been heavily regulated in the United States and Canada for decades, some lakes in Canada have started showing increasing concentrations of these compounds. The cause is not atmospheric transport or illegal use; rather, it comes from melting glaciers. Layers of glaciers that formed in the mid-twentieth

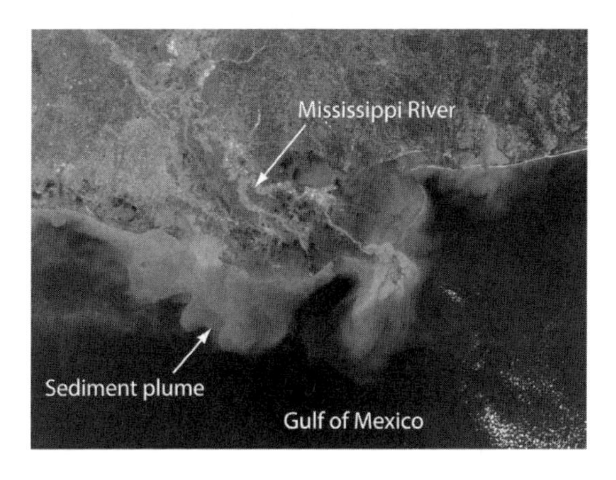


FIGURE 14.1 Plume of Mississippi River water flowing into the Gulf of Mexico. The nutrients and contaminants in this plume can be measured for hundreds of miles and have even been implicated in the demise of Florida's coral reefs. Original satellite image from NASA's Earth Observatory.

century, when glaciers were still gaining mass and DDT and PCBs were commonly used, are now melting and releasing these compounds back into the environment (Blais et al. 2001). The same pattern is occurring in the European Alps, where some alpine lake fish and mussels have DDT in their tissue amounting to more than double the regulatory limit for humans (Bettinetti et al. 2008).

## Changes in Uptake and Toxicity

Climate change can cause organisms to become more sensitive to pollutants, either because of increased general metabolic stress or by affecting the physiologic processes responsible for detoxification. Through changes in the environment, it can also alter the chemistry of pollutants in ways that make them more toxic (fig. 14.2). Many of the factors that affect uptake and toxicity of pollutants, such as pH, salinity, and temperature, are affected by climate change. Warming temperatures can increase metabolic rates, which increases uptake and in some cases reduces an organism's ability to detoxify itself. Reduced salinities, which can be the result of increased terrestrial runoff into coastal marine systems or melting sea ice, can also increase uptake of metals and PAHs.

The effects of pH on chemical toxicity have been studied most extensively in freshwater systems, particularly in terms of acid rain. Reduced pH, or acidification, can result in both enhanced exposure (many metals become more biologically available at lower pH) and enhanced toxicity of pollutants. Today the same increase in atmospheric carbon dioxide responsible for climate change is also causing acidification of the

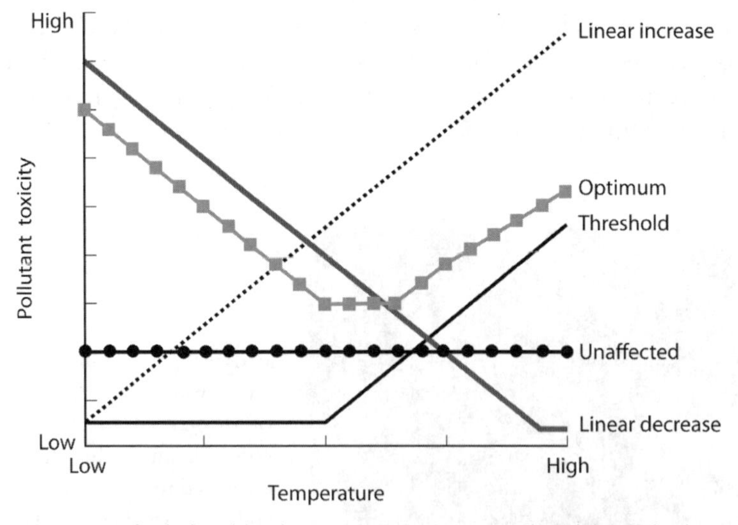


FIGURE 14.2 Types of relationships between temperature and toxicity. Whether toxicity increases, decreases, or remains unaffected by increasing temperature depends on the pollutant in question, the organisms being exposed, and other factors. After Sokolova and Lannig 2008.

world's oceans (see chapter 2). Climate change may also cause acidification in freshwater systems on a smaller scale through increased oxidation of sediments during prolonged drought. It is unclear whether increasing atmospheric carbon dioxide itself is likely to lead to acidification in freshwater systems.

Of course pH is not the only changing environmental variable that will interact with pollutants, nor will the pollutants be alone. Interactions among multiple factors can lead to better or worse effects than anticipated. For instance, the combination of reduced pH and elevated temperature makes exposure to some pesticides more damaging than either stressor alone. On the other hand, elevated temperatures (up to a point) can increase enzymatic repair of damage, leading to overall lower effects for other pollutants. Even though there is less information about these more complex interactions, consideration of their potential effects must be an element of climate-savvy management and regulation, for instance by employing the precautionary principle. This would entail lowering regulatory limits for pollutants to take into account the potential for adverse interactions and the growing pressures that occur as climate change progresses.

Another environmental stressor that can increase toxicity and is likely to be exacerbated by climate change is decreased oxygen levels in freshwater and marine ecosystems (Shaffer et al. 2009). Dissolved oxygen content is in part a function of water temperature, and as water warms, the amount of dissolved oxygen it can hold decreases. Climate change may also reduce oxygen levels by reducing mixing within the water column. Low oxygen levels are already a problem in many coastal areas (fig. 14.3), and climate change will contribute to increasing the size, number, and duration of these "dead zones." While reduced oxygen levels alone are stressful enough for many organisms, they can also increase the time it takes for pollutants to break down into less harmful forms, meaning the pollutants are more dangerous for longer.

FIGURE 14.3 Location and status of oxygen-depleted or dead zones globally. After Nellemann et al. 2008.

### BOX 14.2 DEAD ZONES OF THE WORLD

Dead zones are regions of hypoxic or anoxic (low or no dissolved oxygen) water that lack the diversity and abundance of organisms found in more oxygen-rich surrounding waters. The number, size, and duration of dead zones have increased dramatically since the 1960s, about a decade or so after the onset of widespread use of synthetic nitrogen fertilizer production (Galloway et al. 2008). Increased nitrogen causes algal blooms, which leads to decreased oxygen when the algae die and decay. The same processes that deplete oxygen can also decrease pH, so dead zones often have low pH as well.

Today there are more than 400 identified dead zones around the world, characterized as persistent, annual, or episodic (figure 14.2). They are caused by a combination of factors, most significantly nutrient enrichment from terrestrial runoff, and they can be aggravated by high temperatures and stratification of the water column. Some new dead zones, as well as exacerbation of existing dead zones, seem to be related to current levels of climate change (e.g. Chan et al. 2008), and the prognosis for the future indicates that this is a growing challenge with dead zones projected to be more and more common (Diaz and Rosenberg 2008).

## Incorporating Pollutants into Climate Change Adaptation

While it is well understood that climate change and its effects will influence the toxicity, availability, and use of pollutants, there has been little exploration of how reduction of such stress could fit into climate change adaptation planning. This is an emerging area that is ripe with opportunity and fraught with risk. Just as climate change affects the vulnerability of species to pollutants, pollutants can also affect the vulnerability of species and ecosystems to climate change. Some chemicals make species more sensitive to thermal stress, for instance, putting them at higher risk from the projected increases in average and extreme temperatures across biomes. Some pollutants damage immune function, increasing the risk posed to species (including humans) as disease ranges expand or outbreaks become more common. In addition to these climate-specific increases in vulnerability, organisms that are generally stressed by pollutants are often more vulnerable to any other stress, including changing climate conditions.

Failure to consider how to reduce these interactions will only make systems more vulnerable to climate change. In the Great Barrier Reef Marine Park it has been determined that protecting water quality may have the "most significant impact on resilience" (Marshall and Johnson 2007).

Given the early state of thinking on climate change and contaminant interactions, there are few good, specific examples of how it has been handled successfully. Nonetheless, it is clear that even if we believe they have been dealt with adequately, we should consider pollutant issues in climate change vulnerability assessments and adaptation

### BOX 14.3 CORALS, CLIMATE CHANGE, AND WATER QUALITY

**Vulnerability assessments of the Great Barrier Reef and its associated resources indicate that the combination of poor water quality and climate change adversely affects almost all components of this diverse system. Phytoplankton, sea grasses, corals, fish, dugong, and microbial communities are all affected by some combination of increasing temperatures (or other climate stress) and increased nutrient runoff, sewage, and trace metals or other toxins. Poor water quality increases the risk of coral bleaching. In some near-shore regions with particularly high terrestrial runoff, reducing nutrient pollution could have the same effect on bleaching risk as reducing water temperature by 2 to 2.5°C (Wooldridge 2009). In other words, reducing nutrient pollution might allow coral reefs to withstand an extra 2 degrees or more of warming. The Reef Water Quality Protection Plan created by the Queensland and Australian governments aims to increase the overall resilience of the reef by reducing input of pollutants and restoring natural hydrological features that filter pollutants from runoff. One of the great challenges even this well-studied system faces is the shortage of information on water quality and pollutant loads.**

plans. This means considering both climate change effects on vulnerability to pollutants and pollutant effects on vulnerability to climate change. Some good ways to start adapting pollutant regulation to climate change include:

- Reconsider regulatory limits with the added stresses of climate change included in the equation.
- Reconsider permit requirements. Many permits are contingent on historic conditions (flow rates, seasonal timing, temperature, pH) that may be changing but may not be regularly monitored. More responsive permit rules may be required to factor in altered sensitivity of a system due to climate change and increasing climate variability.
- Refine monitoring and evaluation plans so that interactions among climate change, its effects, and pollution are identified early enough to take action. This is particularly important given the relative dearth of knowledge about how such complex interactions are likely to play out in the real world.
- Create planning or research bodies covering multiple sectors or large geographic scales to assess potential conflicts and synergies between adaptation strategies with regard to contaminant issues. Such bodies can also cooperate to create more effective solutions that avoid problems.
- Use local- or regional-scale climate and contamination scenarios to frame discussion of possible implications for species and habitats. Consider the range of climate-related variables that may have important interactions with pollutants (e.g., pH, salinity, runoff timing, and volume) and focus particularly on

pollutants that are likely to increase vulnerability to climate change. Use these explorations to develop local plans for reducing or managing contaminant and nutrient stress in a climate-savvy manner.

- Consider adjustments to regulatory structures to allow incorporation of uncertainty, including uncertainty about climate change and variability, its effects, the use of toxic compounds, and interactions between climate change and toxics. Variability is quite important to consider as extremes may play a dominant role in determining effect. For example, dissolved oxygen levels may be livable at 4mL/L and, if you consider averages, this may be maintained. However, regulating for an average is irrelevant if conditions drop to a lethal level (say 2 mL/L) for even a few hours occasionally. We will discuss governance issues further in chapter 16.

## Final Thoughts

The combination of multiple pollutants and multiple environmental changes due to climate change and other forces all occurring simultaneously is the sort of complexity that we experience in the real world. While often difficult to parse or to understand, it is nonetheless our reality, and we can call on past experience with managing in the face of complexity to inform future work, now with a new urgency. As with all other aspects of climate change adaptation planning, it is important to think not just about the immediate circumstances but to start thinking also about what physical, chemical, biological, and sociological conditions will be like in the near to distant future. Climate change demands that we build temporal thinking more strongly into our current spatially oriented approach, and that we take a more proactive approach to planning and management. Addressing the multifaceted interactions between pollutant problems and climate change—changes in exposure, toxicity, and transport as well as pollutant influences on climate vulnerability—will be crucial if we hope to keep the environmental gains we have made in this area.

# Chapter 15

# *Integrating the Needs of Nature and People*

We have options, but the past is not one of them.
—*David Sauchyn and Suren Kulshreshtha*

The focus of this book is the conservation and management of natural resources. Yet even practitioners whose focus is not human welfare would benefit from incorporating human needs and uses into their planning. The alternatives—walling off protected areas, policing them vigorously, expecting people to obey laws regardless of their own circumstances, or simply ignoring human concerns altogether—are frequently impractical, expensive, or ineffective. This will be particularly true as climate change decreases the reliability of systems on which people have come to depend. Thus climate change vulnerability assessments and adaptation planning for species, habitats, and ecosystems should consider how humans might respond to climate change or its effects, and how this might influence the vulnerability of natural systems to climate change. Adaptation plans that anticipate and incorporate the opportunities and challenges resulting from human responses are almost certainly more robust than those that do not.

In this chapter we explore approaches to meeting both conservation and human development needs in a changing climate. When do the interests of development and nature mesh? When are they likely to conflict? And how can we use this information to develop more robust adaptation options for both?

## Challenges and Opportunities

Human response to climate change at large and small scales will provide both challenges and opportunities for effective conservation and resource management. Some conflicts or threats will become more pronounced, while others may decrease. By anticipating new pressures, suggesting alternative strategies, and building coalitions before new development and resource use patterns become established, managers and practitioners can save much time, money, and heartache.

### *Water*

One resource for which climate change most clearly contributes to increased conflict is water. Even at current use levels there is frequently not enough water to go around, and the decreasing availability of freshwater projected for many areas will only make the situation worse. This has several implications for conservation and management. There will likely be increasing pressure to build dams and reservoirs to make up for longer droughts or storage capacity lost as glaciers and snowpack shrink. The Okanagan County Public Utility District in Washington State, for instance, is proposing to build a new dam in part to provide a reliable summer water supply as climate change progresses. The reservoir would flood over 18,000 acres, half of them in Canada, in an area that Conservation Northwest refers to as "one of the most biologically, economically, and culturally diverse watersheds in western North America." In addition to flooding previously dry land, such dams radically alter freshwater habitats both up- and downstream of the dams. Recent conflicts over the role of Chinese dams in record-low water levels in the lower Mekong River illustrate the potentially multinational character of the problem. Making strong efforts to decrease water demand and develop alternatives *before* calls for new dams and water diversions arise will help to reduce the vulnerability of nature, agriculture, and cities to drought and variability.

Human responses to climate change may also create new demands for water. The push for cleaner energy sources is stimulating plans for large-scale solar arrays, often in desert environments. Even air-cooled solar power plants require large quantities of water for cleaning solar panels, a real issue in dry regions. The Ivanpah Solar Electric Generating System proposed for construction in California's Mojave Desert will use an estimated 25 to 32 million gallons of water per year (U.S. Bureau of Land Management 2009). The spread of agriculture to higher latitudes may create a greater demand for water in new areas. Again, proactively pushing for alternatives (rooftop solar rather massive solar arrays, for example) can help to reduce new demand, but we all need to incorporate the possibility of unexpected shifts in human behavior and resource use into planning processes.

### *Land Use*

In addition to creating new demands for water in some areas, the push for less carbon-intensive energy sources is creating a new set of land use challenges. In addition to their water needs, massive solar arrays destroy acres of habitat. Wind turbines kill birds and bats. Tropical forests are being clear-cut for biofuel plantations. Yet continuing to rely on coal and other fossil fuels will also destroy habitat, pollute air and water, and, of course, further climate change. In the absence of stabilized or decreased energy demand, new power plants *will* be built, and less carbon-intensive options must be part of the mix. We need thoughtful, well-regulated guidelines for renewable energy development that balance the need for energy and the need to protect natural resources. The Royal Society for the Protection of Birds, for instance, came out in support of expanded wind farm development in the United Kingdom after investigating issues around birds, climate change, and wind farms, and published a report outlining essential elements for environmentally sensitive onshore wind development (Bowyer et al. 2009).

More importantly, there needs to be a much stronger push for increased energy efficiency and for distributed rather than centralized power (for instance, putting solar panels on roofs rather than developing massive solar arrays in pristine desert habitat). Distributed power has the added benefit of reducing the vulnerability of businesses and households to extreme weather events. For instance, Harmony Resort on Saint John, U.S. Virgin Islands, which has solar hot water and solar electric systems, had no loss of power or hot water during hurricanes Marilyn and Georges, while other areas on Saint John and nearby islands suffered utility disruption for weeks or even months (Deering and Thornton 1998).

### BOX 15.1 CLOSING THE LOOP

A shift toward more water-intensive cultivars and farming techniques has increased the vulnerability of agriculture in many regions to climate variability and change. In years with normal rainfall, harvest is high; dry years can reduce production by 50 percent or more (Tan and Reynolds 2003). The problem is compounded by the inefficient irrigation systems common on many farms. Farmers add water, pesticides, and fertilizers to their fields only to lose large quantities to evaporation or runoff, costing the farmers money and polluting waterways. Some farmers are experimenting with closing the loop, installing systems that capture and reuse water from their fields. In Illinois, fields using a closed loop system that collects drainage and runoff from fields for later reuse in irrigation had higher yields than neighboring fields during normal years, and had up to twice the yield of neighboring fields during dry years (Tan and Reynolds 2003 and references therein). Closed-loop systems have the additional advantage of returning sediment and excess nutrients to fields. Some farms take the closed-loop idea even further, feeding vegetable waste from farming operations to dairy cows, then converting the resulting manure into either liquid fertilizer for their fields or natural gas to help power the farm. For the farmer, the benefits are reduced topsoil loss, higher yields, and lower costs for water, fertilizer, and energy. For ecosystems, the benefits are reduced sediment load and nutrient pollution. And all parties are more likely to get the water they need.

Responses in the agricultural sector to climate change are another important determinant of future land use patterns. There may be a push to develop agriculture at higher latitudes, for instance, or farmers and herders may abandon areas where climate change has made their lifestyle no longer tenable. Herders in regions as disparate as northern Sweden and Mongolia, for instance, are losing increasing numbers of livestock to harsh winter conditions that may be linked to climate change, and farmers in some dry areas of Madagascar are moving to the coast to take up fishing. Although there are efforts to genetically engineer crops for projected future conditions, it is not clear that this will be more effective than traditional agricultural breeding or crop diversification. From a climate vulnerability (and biodiversity) perspective, agricultural systems that depend heavily on a small number of crops or varieties are worse than more diverse systems. They may maximize yields under one set of climate conditions, but leave farmers and consumers with few options when conditions shift due to climate variability or change. Again, the key is to anticipate and be ready for new challenges so that conservation and management strategies are more robust to a range of potential futures.

### *Development*

Sea-level rise, more frequent and severe extreme weather events, and increased coastal and inland flooding may lead to stronger pressure for shoreline hardening or other defensive structures, actions that increase the vulnerability of natural systems. But as the cost of supporting development in hazard-prone areas increases, the willingness of governments, insurers, and others to continue to do so may diminish, creating an opportunity to enact rules and regulations that reduce hazard risk for both people and nature. The value of flood- or erosion-prone land may also decrease, making it more affordable for governments and land conservancies to purchase it for conservation or other purposes.

Population densities will change in response to a range of factors. In addition to movement as a result of increased costs of living along dynamic coasts or floodplains, people will also move in response to livelihood options. As mentioned previously, some inland farmers in Madagascar are moving to the coast to pursue fishing as increasingly dry conditions make farming less tenable. Logging communities may shrink in areas increasingly devastated by bark beetles, fire, or other climate-related factors, while communities may grow in areas where a lengthening growing season makes farming more profitable. A shift toward lower-carbon energy sources may force people currently employed in the fossil fuel sector to move in search of other jobs, perhaps to areas rich in renewables rather than in coal or oil. Establishing development plans that anticipate or promote the departure or influx of groups of people may steer community growth in more sustainable directions.

### *Invasive Species*

As discussed in chapter 12, climate-related shifts in invasive species will be linked to direct effects of climate change as well as human responses to it, with potentially significant implications for both conservation and human welfare. The expansion of pests and diseases already provides a challenge for people and wildlife alike. While such expansion is difficult to stop, it can be anticipated through monitoring, education, quarantine, establishment of early-response plans, and crop or livelihood diversification.

Particular challenges come in the form of well-intentioned species introductions as responses to climate change. The perceived profitability or environmental benefits of biofuels may lead to the introduction of invasive nonnative species to new areas, or conservation-oriented individuals may turn to assisted migration to help species they perceive as threatened by climate change. While noble from the perspective of fighting climate change or saving endangered species, the introduction of nonnative species to new habitats is fraught with risks. Managers of potential recipient ecosystems should keep an eye out for efforts to bring "climate refugee" or biofuel species to their area, and ensure that such decisions reflect the welfare of the recipient ecosystems as well as the welfare of the potential refugee species or the biofuel industry. This is not to say that purposeful introduction of species beyond their native range should be excluded from the adaptation or mitigation toolbox, simply that it must be undertaken only after a careful assessment of possible environmental impacts.

## Making Existing Tools Climate Savvy

There are many existing tools and best management practices geared toward making planning and development less environmentally damaging and less vulnerable to natural hazards. This same suite of options can be employed to address the effects of climate change. As is the case under current climate conditions, areas with strong, thoughtful planning processes are generally less vulnerable to natural hazards in the face of climate change. Here we briefly review options for making existing tools more climate savvy.

### *Comprehensive, Land Use, or Other Community Plans*

Communities and organizations develop plans at a variety of scales, from neighborhood to municipality to coastal zone to continent. Some focus on individual issues (e.g., post-disaster recovery), while others are broader. The Washington State Growth Management Act, for instance, mandates that the state's fastest-growing counties develop comprehensive plans encompassing land use, transportation, housing, economic development, environmental protection, public facilities and services, historic lands and buildings, shoreline management, and property rights. A strong comprehensive plan or a history of strong planning in general makes adaptation much easier as it provides a readymade context for considering values and trade-offs among various interests, and a structure that can be built upon to reduce vulnerability to climate change.

One step in updating comprehensive plans or zoning is to update maps of natural hazards, such as mudslides or drought, and critical areas, such as wetlands or aquifer recharge areas, to reflect projected changes. This allows development to be sited appropriately for future risks, reducing the vulnerability of human communities as well as

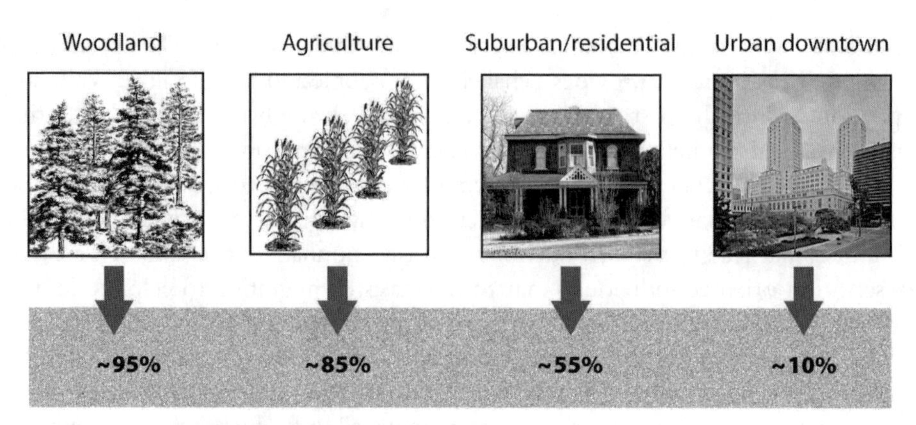


FIGURE 15.1 Rough percentage of water percolating into the ground as a function of land use. These percentages are representative only; actual percentage of water soaking in versus running off depends on variables such as soil type, lot size, and so on.

giving nature more room to adjust to changing conditions. Maps of projected water availability may help direct development away from areas where water availability is likely to become limiting, and help policymakers promote land use designations that support adequate recharge levels (fig. 15.1). Using updated maps, planners can make better use of the tools in their toolbox, such as buffer areas, setbacks, clustered development, and planned retreat. In areas with rapid urbanization and growth, such as much of Asia, comprehensive planning is both more difficult and more important.

An excellent example of longer-term planning comes from the United Kingdom's 2006 Shoreline Management Plan (SMP) guidelines (DEFRA 2006). The guidance states that planners need to "assess and manage flooding risks and coastal erosions over a consistent timescale (100 years)," and be aware of "longer-term implications (50 to 100 years) of coastal change, climate change, and rises in sea levels." The guidance was piloted in three different coastal regions, and the resulting SMPs illustrate the range of approaches taken. Where coastal defenses protect valuable assets that cannot be easily moved (e.g., large urban areas), SMPs typically call for "holding the line" (maintaining defenses) over both the short and long term. In other cases, holding the line is advocated only in the short and medium term, while assets are relocated. The route taken reflects a range of considerations, including effects on other sections of coastline. In the case of one gas terminal, for instance, coastal defenses blocked up to 70 percent of sediment input for coastal areas down-current of the site. The recommendation was to maintain the gas terminal for the remainder of its operational life, roughly fifty years, at which point the terminal would be decommissioned rather than replaced.

### *Regulatory Tools*

As mentioned above, regulatory tools such as zoning and permitting can be made more climate savvy by updating hazard maps and establishing time horizons over which decisions should be made. In Australia, there are a number of precedents for using a 100-year time horizon in considering development permits in coastal zones, as in a refusal by the Victorian Civil and Administrative Tribunal to allow building in an area close to the coast because of the risk of flooding and inundation due to sea-level rise over the next century. While many jurisdictions use a shorter time horizon for planning (a thirty-year horizon is common in the United States), such short time horizons do not accurately reflect the typical life span of homes, buildings, and other infrastructure.

Other factors to consider in updating regulatory tools for a changing climate include flood and drought risk and storm frequency and intensity. One possibility is to systematically review common tools and guidelines with an eye for how, if at all, they could be adjusted. In Nova Scotia, a consortium of governmental agencies, private companies, and nonprofit organizations developed a draft guide for incorporating climate change into the Environmental Impact Assessment (EIA) process, and tested these guidelines against actual projects. Adjustments in the EIA guidelines reflected both changes in the vulnerability of projects to environmental factors and of the environment to project activities as a result of climate change.

Regulations typically require special consideration of sensitive or critical habitats in development, permitting, and planning decisions. Another element of making regulations more climate savvy, then, is to evaluate how the importance and sensitivity of particular habitats may change as the climate changes. What is considered an acceptable level of deforestation may change in hilly areas where more intense rainstorms are expected, since the importance of vegetation in preventing erosion and landslides on steep terrain will increase. Wetlands may become increasingly important for their role in flood protection, thereby increasing the weight given to protecting them in situ. The location of habitat critical for the survival of endangered species may shift over time. Ensuring that rules and regulations continue to provide at least their current level of environmental protection requires that we evaluate how climate change might affect various components of those rules and regulations.

### *Infrastructure, Public Facilities, and Capital Improvement*

Performance standards, design criteria, and assessment protocols can also be updated to decrease the vulnerability of infrastructure to climate change as well as the negative effects of infrastructure on the climate vulnerability of ecosystems and species. At a minimum, standards and criteria should be regularly updated to reflect projected future conditions. For instance, the Q-100, or 100-year flood levels, are a standard element in many floodplain regulations, and clearly will be changing with climate change. New bridges should be high enough and strong enough to withstand the largest expected flood over their life span. Required setbacks along coastlines should reflect projected local rates of sea-level rise.

Some practices that are already best practices under current conditions become even more important in the face of climate change. In areas where intense rainfall events are likely to increase, increasing soil infiltration rates through the use of permeable pavement and other low-impact development techniques becomes even more

**BOX 15.2 PERMEABLE PAVEMENTS**

Permeable pavements can provide the benefits of pavement while allowing more water to filter through to the soil and groundwater underneath. Possible materials include porous concrete, porous asphalt, paving stones, and bricks. In addition to reducing both drought and flooding, permeable pavements reduce the amount of pollutants entering the groundwater and the amount of sediment in surface water. In areas where sewage and stormwater systems are combined, permeable pavements also reduce the frequency of sewage overflow events during heavy rain, reducing stormwater system costs and negative public health effects. Permeable pavements can even provide social benefits: in Philadelphia, public basketball courts with permeable pavement are popular because they don't flood during rainstorms. Permeable pavements do have a lower load-bearing capacity than traditional pavements, but work well in parking lots, pedestrian areas, and other places with low or moderate traffic volume. The up-front cost of porous pavement is sometimes but not always higher. Any excess costs are typically more than offset on a project scale by the reduced need for runoff-related infrastructure such as retention ponds, and on a municipal level by reduced flooding or combined sewer overflow events.

cost-effective. Stronger guidelines limiting public investments that encourage or support development in high-risk areas likewise do even more to reduce the vulnerability of people and ecosystems.

### *Increase Awareness and Capacity of Planners, Designers, and Developers*

In many cases, planners, designers, and developers may not be fully aware of the implications of climate change for their work, the existence of tools that facilitate climate-savvy planning, or approaches to limiting vulnerability to climate change. Providing information and guidance on impacts, tools, and approaches increases the likelihood that climate-savvy, environmentally appropriate actions are taken. Building and demonstrating public support for such approaches may also reduce barriers to their uptake.

Awareness-building and support can take many forms. Convening workshops in which planners, conservation practitioners, and resource managers receive training together may build community and mutual understanding as well as adaptation-specific capacity. This may be particularly true for GIS-based tools that allow easy mapping and visualization along with data analysis. The Nonpoint Source Pollution and Erosion Comparison Tool (Coastal Services Center 2004), for instance, allows users to investigate the water quality impacts of land use, development, climate change, or interactions among these factors, and could help all users better visualize possible futures for their region.

Another approach is to provide managers and practitioners with case histories illustrating how others in similar situations have built climate change into planning and development without sacrificing environmental integrity and vice versa. This could be done via speaker series or using the rapidly growing number of published or online collections of climate adaptation case studies.

## Motivating People to Act

Climate change does not typically raise new questions or dilemmas when it comes to balancing the needs of development and ecosystems: it simply changes the calculus. Thus many existing skills and questions remain important. In particular, good negotiating, politicking, and community organizing skills, which have been central to past conservation and management successes, will likewise be central to the success of adapting to climate change. Actions must be calibrated to the audience in question. If people remain convinced that climate-related hazards are unimportant compared to other concerns, further education about the scope and seriousness of climate change may be important. On the other hand, if people's resistance comes from difficulty with facing up to large, imminent problems, taking an alarmist or "climaggedon" approach only makes things worse. As general awareness of climate change grows, generic information about its effects may become beside the point or even frustrating to practitioners who are instead looking for targeted information and guidance to help design strategies given the time, money, expertise, and political structure at their disposal. In these cases it may be best to approach the issue in the context of familiar cost-benefit, planning, and regulatory systems.

## Resistance versus Resilience

An overarching theme of natural hazard planning is the balance between focusing on resistance to hazards or on the ability to recover quickly (resilience). The former relies more on strength, the latter on flexibility. Disaster management headquarters should likely be designed as fortresses able to withstand the strongest storm or flood. In contrast, a resilience-based approach might be better for less critical infrastructure or for critical infrastructure in areas where a fortress approach is not feasible. The mobile bathhouses and clay-and-shell roads in Maryland's Assateague Island National Park, discussed in chapter 6, are a solid example of resilient infrastructure. The shift from asphalt to clay-and-shell roads has the added benefit of minimizing repair costs following storms and flooding.

While a resistance-based approach may be essential in some cases, in others it can increase vulnerability by creating a false sense of security. Once some defenses are in place—levees or other flood-control structures, for instance—people begin to view hazardous locations as safe. As more people move to hazard-prone areas such as floodplains or dynamic coastlines, more lives and infrastructure are exposed to the natural hazards characteristic of those locations. And while resilience in one realm—social, economic, infrastructural, or natural—is generally linked to resilience in others, the same is not true when it comes to resistance. Human communities that take a resistance-based approach typically reduce the resistance and resilience of the natural communities around them. Unfortunately, federal governments and international aid agencies have typically been more likely to support resistance-based activities such as shoreline armoring or beach nourishment than to provide support for the sort of strong local planning that could facilitate a resilience-based approach.

By pushing for more thoughtful analysis of when resilience-based approaches to development and infrastructure may be more appropriate, and by helping to develop resilience-based options, conservation and natural resource managers may help to decrease the vulnerability of both nature and humans to climate change.

## Final Thoughts

The tensions between conservation and development, and between short- and long-term benefits, will doubtless never be resolved. There are win-win situations, but there are also situations where trade-offs cannot be avoided. Climate change affects the relative costs and benefits of various actions, so smart planners will build this reality into their planning process at many levels. People will not stop trying to protect themselves and their property from natural hazards, so conservation and resource management practitioners must offer options that reduce the vulnerability of the natural world without increasing that of human systems.

# Chapter 16

## *Adapting Governance for Change*

You can't change the science, you can change the politics.
—*Hans Verolme*

Over time, communities develop institutions and processes for making decisions, setting policies, or sharing power that work for their particular circumstances. When social, economic, technological, or ecological conditions are relatively stable, rigid governance structures can work well, allowing sustainable use of natural resources for decades or centuries. When conditions change rapidly, however, rigid governance structures frequently weaken or fail. To govern and manage effectively in the face of rapid change, we need legal and regulatory mechanisms that facilitate responsive, effective conservation and management.

The concept of adaptive governance has arisen in response to this need (e.g., Dietz et al. 2003; Folke et al. 2005). While still evolving, the overall goal is to create governance systems that balance flexibility with oversight; discretion with accountability; science with politics. This does not always require new laws; in some cases it is sufficient to change the interpretation or implementation of existing laws. As Kenneth Kakura says, "If you can't change the law, change the practice."

Adaptive governance and its implementation through adaptive management are distinct concepts from climate change adaptation. Adaptive governance and adaptive management can be part of effective adaptation, but neither technique is adaptation in and of itself. Climate change adaptation refers to reducing vulnerability to climate change specifically; adaptive governance is one mechanism that allows it to happen.

## Thinking Ahead

Environmental laws, regulations, and management plans tend to be created in response to observed rather than anticipated problems, and are rarely designed with an eye toward the possibility of unanticipated problems. In this way they are fighting fires rather than preventing or managing the conditions that cause or feed fires. They respond reactively, in the same manner they were created.

Such an approach is inappropriate in the face of rapid changes in climate or other conditions. Climate change adaptation is ideally built around a proactive vision for the short, medium, and long term that accounts for complex scenarios and interactions that go beyond the direct effects of climate change itself. To capture and address this complexity, information and infrastructure must be provided by a diversity of players at a diversity of scales. While immediate needs must be met, a key function of adaptive governance structures is to make sure that short-term benefits are not always given priority over long-term stability.

## Holistic Planning

Climate change has wide-ranging effects across multiple sectors, and interacts with multiple stressors simultaneously. In contrast, many governance and regulatory structures deal with different sectors and stressors independently, and solutions for different sectors, habitats, or threats are often developed in isolation. Despite a multitude of ecological land-sea connections, for instance, management of marine and terrestrial resources is typically not coordinated. Many coastal states address sea-level rise independently of other climate change challenges that will accompany or compound it. Transportation departments may develop adaptation strategies separate from those developed by public lands departments. The result could be conflicting plans or priorities for the same land and water resources.

Cross-sectoral governance and regulation can help to address these issues and to maximize benefits and minimize risks of adaptation efforts. Cross-sector awareness or engagement means that plans are not developed for a single effect (e.g., sea-level rise, reduced precipitation, increasing temperature) or for a single sector (e.g., forestry, agriculture, recreation) in isolation. To make integrated planning a reality, we may need governance structures and regulatory mechanisms that can reach across sectors, or at least foster coordination among them. For example, the United States' National Environmental Policy Act (NEPA) requires that all federal agencies use a "systematic and interdisciplinary approach" to evaluating environmental implications of their planning and decision making activities. Agencies must assess both imminent and longer-term effects, making NEPA temporally as well as sectorally holistic. Many coastal zone management laws around the globe also support holistic governance, providing a mechanism for a range of interests and organizations to work in coordination with territorial or federal agencies to manage coastal resources for economic development and environmental protection.

Some countries are experimenting with ways to develop holistic planning on an even grander scale. Bhutan's development philosophy of Gross National Happiness is built around four main pillars: environmental conservation, equitable economic development, cultural promotion, and good governance. With the input of the Gross National Happiness Commission, the goal is to mainstream these concepts into policy development and implementation. Tangible objectives—for instance, the preservation in perpetuity of 60 percent of land cover as forest—were set for each pillar (Kingdom of Bhutan 2008). In Ecuador and Bolivia the national constitutions include the indigenous Quichua people's concept of *buen vivir*, or good living, which includes an implicit interest in promoting human well-being while living in harmony with nature. Ecuador's constitution explicitly recognizes Pacha Mama (loosely meaning Mother Earth) as an entity with fundamental rights, and guarantees its citizens the right to an ecologically balanced environment. In theory, these philosophies encourage actions that are widely beneficial rather than primarily benefiting special interests or a relatively small wealthy class. Interpreting these governance frameworks through a climate-change lens could mean that adaptation strategies that provide broad, long-term benefits receive priority over those that address issues in isolation and only over the short term. The developing concept of intergenerational planning and justice could also be applied to these challenges (Weston and Bach 2009).

## Stakeholder Engagement

When it comes to successful long-term governance of common-pool resources such as fisheries or forests, the existence of strong community networks and resource users' support for monitoring and enforcement is essential. These elements become even more crucial if conditions are changing rapidly, because regular resource users and community members can observe and interpret changes on a scale that formal or external monitoring may not capture. The collapse of the northern cod fishery in Atlantic Canada, for instance, may have come about in part because scientists did not give sufficient weight to the concerns of inshore fishermen (Dietz et al. 2003). Furthermore, social pressure from an engaged community can sometimes be a cheaper and more effective way to get people to follow rules and regulations than top-down, centralized enforcement.

Stakeholder engagement is not necessarily limited by political boundaries. In the Amazon, the tri-national MAP Initiative arose in response to the development of the Inter-Oceanic Highway in an effort to minimize the negative effects of this massive road-building project. The initiative provides a structure for individuals, communities, and governmental and nongovernmental organizations to discuss concerns and promote "environmental governance" (Perz et al. 2008). A series of fora and working groups have been developed, including around the issue of climate change, to build constituents' capacity to play an active and informed role in environmental governance. While the focus of the initiative is road-building, the issues it addresses are intimately linked with climate change in multiple ways, and the initiative has created a mechanism whereby those reliant on forest resources can make their voices heard at all levels of government.

## Integration of New Information

New information relevant to policy and management are constantly becoming available, including information about climate change, its effects, and effective responses to it. There must be ways to integrate this information both into existing plans and processes and into new adaptation strategies. This can take many forms, including specified triggers for reevaluation based on monitoring plans or regularly scheduled annual updates and revisions to plans, processes, or regulatory limits.

The South Bay Salt Pond Restoration Project in California's San Francisco Bay illustrates one approach to this issue. The Final Environmental Impact Statement/Report (EIS/R) evaluates project alternatives over a fifty-year planning horizon and explicitly includes climate change effects on habitats, species, and flood hazards (U.S. Fish and Wildlife Service and California Department of Fish and Game 2007). It outlines a mechanism for incorporating updated sea-level rise estimates at the design stage of each new phase of the project, incorporating marsh accretion rates and land subsidence due to water withdrawal from aquifers as well as global changes in sea level. The EIS/R also includes an extensive adaptive management plan that lays out key ecological and social uncertainties related to the project as well as the monitoring, applied studies, and modeling necessary to address those uncertainties. In particular, the adaptive management plan establishes specific triggers—indicators that the project is not on track to reach a restoration target—that would signal managers to assess the situation and take corrective action if needed. Thus monitoring is designed to serve two functions: increasing understanding of the system as a whole and tracking project success. The adaptive management plan even addresses the institutional structures and procedures needed for its successful implementation.

Another example of establishing triggers for action based on possible effects of climate change comes from the Great Barrier Reef Marine Park Authority (fig. 16.1). Under this model, different environmental conditions beget different management decisions that in turn can inform governance decisions. As environmental conditions worsen for the reef—corals experience heat, bleaching, and mortality—the management and governance needed to address it go from local protection to regional planning that incorporates human adaptive capacity.

## Flexibility

Many environmental laws and implementation frameworks are designed for a given set of conditions that are assumed to be relatively static. Opportunities for review may be provided on only a multiyear, sometimes decadal, basis with little opportunity for

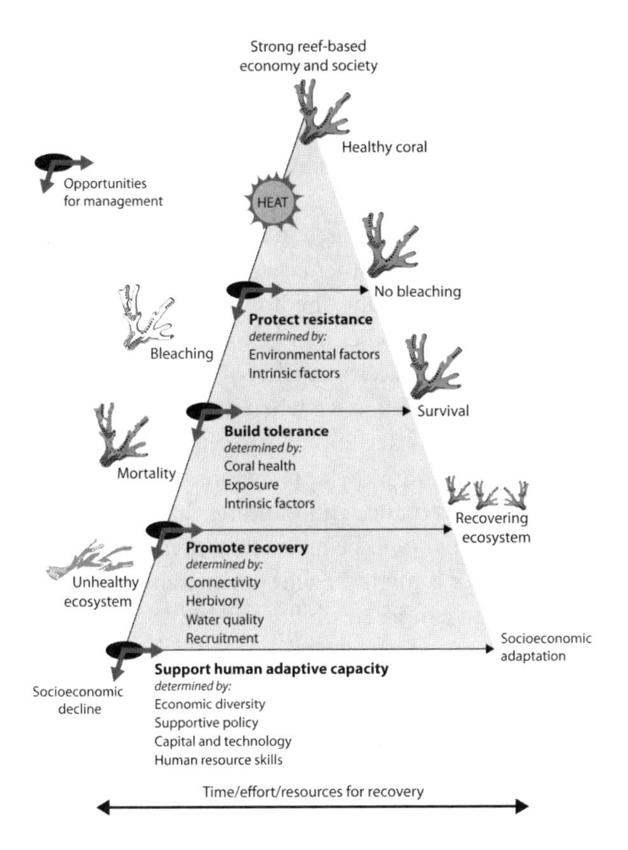


FIGURE 16.1 Opportunities for management intervention to reduce social and economic damage from coral bleaching on Australia's Great Barrier Reef. Increasing temperatures cause bleaching and mortality of corals, but the ultimate ecological and socioeconomic outcome can be influenced by management action. From Marshall and Shuttenberg 2006.

change in interim years. The situation is compounded where laws are designed for top-down command-and-control regulation, which can make timely assessment of and response to local changes even more difficult and expensive. This approach may have seemed appropriate with our past belief that change and variability in the natural world were fairly low, but we may need more responsive mechanisms under conditions of rapid climate change.

The progressive nature of climate change, as well as its tendency to increase climatic variability and extremes, means that governance and regulatory structures need to be flexible enough to adjust to changing circumstances, yet firm enough to maintain their integrity in meeting the underlying challenge (water quality, public health, managed development). A good example is the United States' Clean Water Act, which sets regulatory limits for water quality. The act includes mechanisms for flexibility in the implementation of the law and for developing the law itself. It allows seasonal adjustments in the permitted pollutant levels for compounds whose toxicity changes with changes in temperature. It also includes a contingency planning approach to deal with changes in effluent release, water flow, or additional factors that degrade water quality. This approach requires monitoring to determine whether water quality goals are being met and to support adjustments once deviances are identified. The act itself has been modified over time with amendments addressing shortcomings in the original legislation's ability to meet its goals. Through these changes the act has remained effective and relevant despite changing political, economic, technological, and climatic circumstances. The United States' Safe Drinking Water Act takes a similar approach, allowing revisions to the maximum allowable level of a pollutant as new data become available. The key with climate change will be making sure that data addressing interactions between climate change and toxicity of pollutants are generated in a timely fashion.

Some traditional governance approaches provide other possible models for adaptive governance. Tano (2006) suggests that maintaining the holistic approach traditionally taken by Native American tribal governing bodies may be more effective in the face of climate change than creating new climate policies outside of existing systems. He states that tribal "climate policies will be more effective when they are embedded within broader strategies designed to make tribal development paths more socially, economically, environmentally, culturally and politically appropriate." Tano further cites the historic agility of Native American governance and indicates that some of its features—meaningful and integrated institutions—will be useful in addressing climate change. No governance model is a panacea, but it might benefit policy framers to consider aspects of traditional governance that give more flexibility and holistic structure than is common today. Indeed, building a varied system of multilevel institutional structures such as combining traditional and modern approaches may be an essential element of adaptive governance (Folke et al. 2005). Having a mix of institutional types with a range of decision rules means that information (both scientific and social) will be provided and analyzed from many perspectives. If appropriately implemented, this blended framework may help to provide the responsiveness of a bottom-up approach with the larger vision of a top-down approach.

## Tools for Managing under Uncertainty

There is no universally understood "best approach" to management or governance: what is appropriate depends on the circumstances (fig. 16.2). If uncertainty is low and important determinants of future outcomes can be controlled, an optimal control approach such as setting maximum sustainable yields for fisheries makes sense. In the case of climate change, uncertainty is high: while not unprecedented in the history of the world, the current rate of change is unprecedented in the history of modern humans. Optimal control strategies are therefore much riskier, because our expectations for how the future will unfold are more likely to be incorrect. Here we explore three common approaches to incorporating uncertainty into governance and management: the precautionary principle, adaptive management, and scenario planning.

### *The Precautionary Principle*

When the future is uncertain, many people invoke the precautionary principle to avoid acting (or not acting) in ways that we may regret later. This means that if a behavior

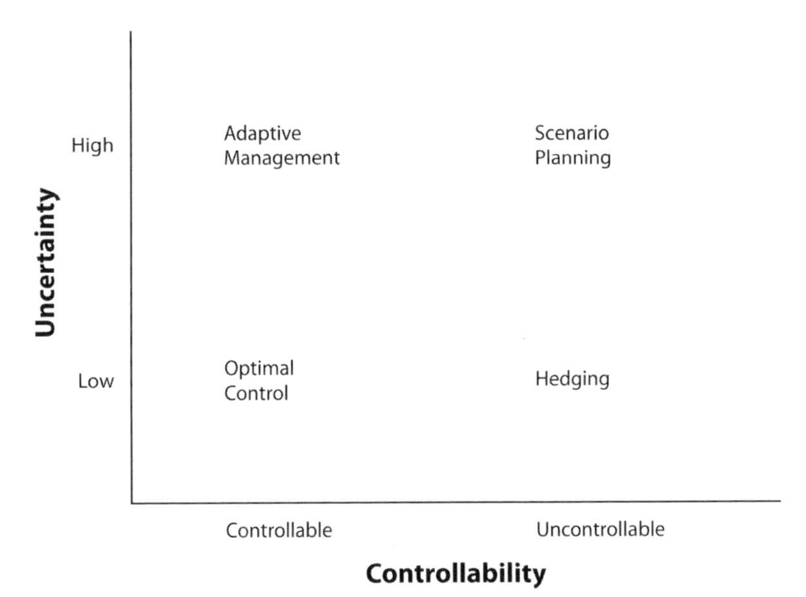


FIGURE 16.2 Appropriate management approaches as a function of uncertainty and controllability (the degree to which managers can control sources of uncertainty). When uncertainty is high, scenario planning and adaptive management are more appropriate. After Peterson et al. 2003b.

(e.g., smoking) or set of conditions (e.g., climatic change) may harm people or the environment, we should try to avoid it, even if we are uncertain of the exact nature or severity of the adverse outcome. The precautionary principle is what motivates people to buy insurance for their home or car. It is not known whether there will be a fire or collision, but insurance offers a degree of protection should either come to pass.

The idea can be illustrated conceptually using the diagrams in figure 16.3. The horizontal axis represents our predictions about how severe some sort of event or change (e.g., drought frequency or duration) is likely to be. If the predicted severity is low, there is no need to act, but as the predicted severity increases, it eventually crosses some threshold (represented here by the vertical line) beyond which we would take action to prevent undesirable consequences. The vertical axis reflects the actual severity of the event or change, and the horizontal line the threshold beyond which undesirable consequences would occur should we fail to act.

The four sections of the graph in figure 16.3a represent four possible outcomes: positive, negative, false positive, and false negative. In the upper right is the so-called true positive. We predicted that the severity of the event would be high, took action to prevent undesirable consequences, and those actions turned out to be warranted (event severity was in fact high). This would be as if meteorologists predicted severe drought, water conservation strategies were enacted, and a severe drought did in fact happen. In the case of false positive outcomes, the predicted severity was high, we took action, and the event or change was not all that severe. Using the drought example, this would be as if meteorologists predicted a severe drought and water conservation strategies were enacted, but the drought never happened.

In contrast, a negative outcome (lower left) results when the predicted severity was low, no action was taken, and the actual severity was also low. Note that the term negative here does not mean that the outcome itself was undesirable—it is a good thing if there is no drought. It simply means that both the predicted and actual severity of the event or change were low. A false negative outcome *is* bad. This happens when the predicted severity was low so we took no action, but the actual severity was high enough that we suffer undesirable consequences of our inaction. This would be as if meteorologists predicted no drought, people made no effort to conserve water, and a severe drought led to water shortages.

Because we cannot know ahead of time how good our predictions are, we have to decide whether we would rather risk not acting when we should have (false negative) or taking unnecessary action (false positive). If the cost of a false negative is relatively high (e.g., water conservation is more costly than water shortage), we should shift the action threshold to decrease the chance of a false negative (fig. 16.3b). Alternatively, if the cost of a false positive is higher (water conservation is cheaper than the possible effects of the drought), the opposite would be true (fig. 16.3c). This is in essence the precautionary principle in action.

Figure 16.3d illustrates how this might play out relative to preparing for the potential increase in droughts projected for many regions as a result of climate change. Decision makers can enact policy and infrastructure changes in anticipation of more droughts—provide no-interest loans for industries to shift to less water-intensive technologies, mandate low-flow toilets and showerheads, raise the cost of water, and so on—or they can wait until something happens. Advance action might be costly, but there are few remedial actions that can be taken once the drought starts. The four possible outcomes are:

1. Both water conservation and drought frequency increase (true positive). The investment of money and political capital averted a water crisis.
2. Water conservation increases, but drought frequency does not (false positive). From the perspective of changing drought frequency, the investment of money and political capital was not warranted.
3. No action is taken, but there is a drought (false negative). No money or political capital was invested ahead of time, and the region must bear the medical, social, environmental, and economic cost of not having enough water.
4. No action is taken, but there is no drought (true negative). No money or political capital was invested, and none was needed.

The potential costs of a false positive or false negative should be evaluated more broadly than simply in terms of projected increases in droughts. Decision makers must also consider potential benefits of water conservation regardless of any change in drought frequency. Many regions are already facing water shortages, and those that are

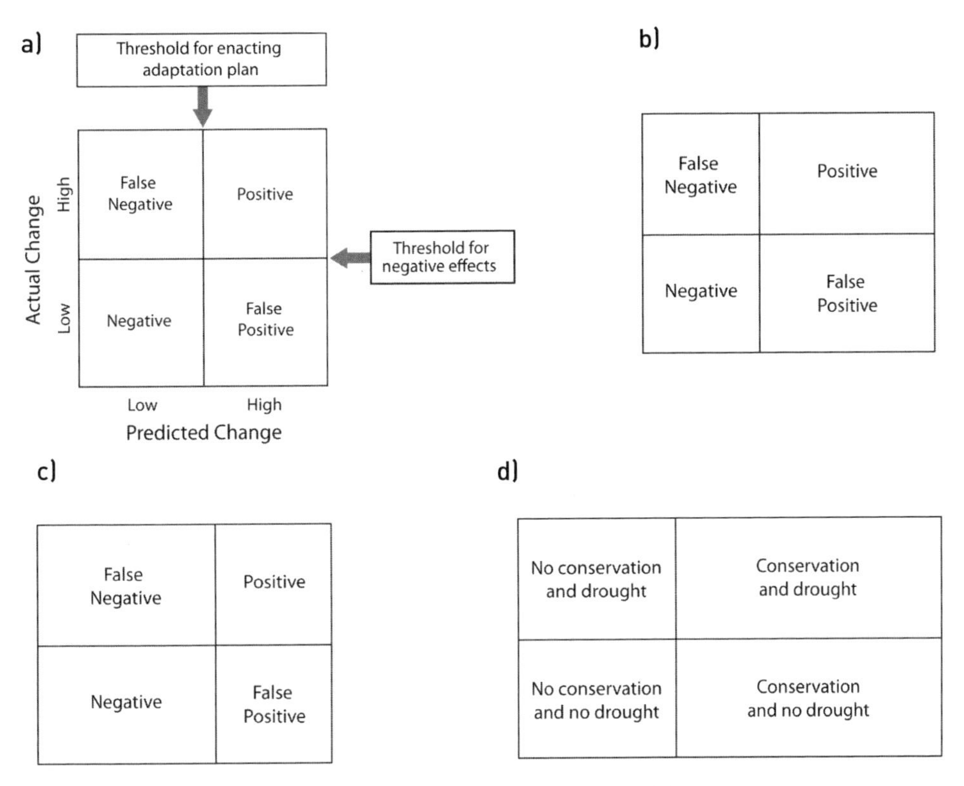


FIGURE 16.3 Thresholds for action and the precautionary principle. General concept (a); action threshold adjusted to decrease risk of false negative (b); action threshold adjusted to decrease risk of false positive (c); potential outcomes for responses under drought uncertainty (d).

not may do so in the future regardless of drought if water supply remains the same but the number of people using that water increases. Enacting water conservation measures may lower water bills or decrease the costs of water infrastructure development and maintenance.

The precautionary principle is easier to apply when costs and benefits of action or inaction are shared equally, or when all parties agree on the relative costs or benefits of different actions. Unfortunately, this is not always the case. Shutting down commercial fishing in a particular region to avoid the risk of overharvest, for instance, would be costly to large fishing corporations but beneficial to subsistence or recreational fishers.

## *Adaptive Management*

Adaptive management is a hypothesis-driven, experimental approach to natural resource management and conservation. While most management structures would allow change if it became clear that current strategies were not working, adaptive management takes a conscious and active approach to generating and testing key hypotheses about the relative effectiveness of different management approaches.

The first step in adaptive management is to define the goal or purpose of the management system in question, for instance balancing harvest with sustainable population levels or maintaining the health of a popular coral reef dive site. With this goal in mind, explicitly lay out the current understanding of how the system works, highlighting uncertainties and unknowns that might affect management plans, and develop possible management plans that incorporate both what is known and what is not. During this process, identify the kinds of information you will need in order to evaluate your understanding of the system as a whole as well as the effectiveness of different management plans. Use this to develop a monitoring plan as well as mechanisms for feeding information from that plan back into management decisions. The importance of well-designed and well-executed monitoring cannot be overstated: without it there is no way to test assumptions or evaluate management effectiveness. If monitoring plans are not carried out, the data are not analyzed regularly, or decision makers do not use the results of the analysis to adjust management plans as needed, then adaptive management is not happening.

A classic example of adaptive management is the process for regulating mallard duck harvest in north-central North America (Nichols et al. 2007). Managers in Canada and the United States agreed on program objectives, then developed a set of modeling, management, and monitoring plans based on those objectives. The management alternatives consist of three regulation "packages" based on liberal, moderate, and conservative harvest levels. Each package specifies a particular set of daily harvest limits and season lengths for the four waterfowl flyways (roughly equivalent to administrative units for duck harvest) in the region. Every August, before the start of the hunting season, managers decide which package to use for the year ahead. To make this decision, they rely on information from the models and the monitoring program.

Because of uncertainty about the effect of harvest levels on overall annual mortality rates as well as the effects of population density on mallard reproductive success, managers use four different models to project optimal harvest rates each year. The results from these four models are combined to reach a decision about the most likely optimal harvest rates. Models are run each year before the hunting season begins to inform the annual decision about which management alternative to use, then again after the hunting season to determine how well the different models worked. Likewise, data on mallard population levels are gathered each year so that models and harvest decisions are made using both historical and current information.

Although the above example does not explicitly address climate change, the management structure can accommodate climate change in several ways. The annual use and assessment of four distinct population models increases opportunities to find and incorporate any changes in population trajectories due to climate change. Having an established annual schedule for data collection, analysis, and management response prepares both managers and resource users for rapid response to changes in mallard populations. Adjusting the system to address climate change more directly could be accomplished by incorporating climatic variables into the annual analysis and modeling.

Although the frequently complex web of national and subnational environmental regulations can act as a barrier to adaptive management, there are options for successful adaptive management even within existing contexts. For instance, the United States' National Environmental Policy Act (NEPA) requires that an Environmental Impact Statement be prepared disclosing possible environmental effects for any major federal action. Once a project is under way, a Supplemental EIS (SEIS) is required if "[t]he agency makes changes in the proposed action that are relevant to environmental concerns; or [t]here are significant new circumstances or information relevant to environmental concerns and bearing on the proposed action or its impacts" (Code of Federal Regulations 2005). While this increases the regulatory costs of changing course, it also provides a safeguard against managers making environmentally unsound changes to a plan.

Agencies can decrease the likelihood of needing to prepare an SEIS by including a range of possible management adaptations in the initial EIS. As explained by Williams et al. (2009):

> If management adaptations that could occur in light of new information are fully documented and analyzed at the beginning of a NEPA process, the need to supplement NEPA documents may be reduced. Put differently, if an EIS anticipates significant information that can arise from monitoring and assessment, the agency may not need to supplement the EIS when invoking management changes based on the newly acquired information.

This approach has the added benefit of creating a more rigorous adaptive management planning process, since those engaged in developing the plan must systematically consider what management changes they might make based on different monitoring results or other contingencies.

### *Scenario Planning*

In the words of Peterson et al. (2003b), scenario planning is "a systemic method for thinking creatively about possible complex and uncertain futures." Originally developed by military strategists in the U.S. Navy after World War II as a means of anticipating and preparing for a range of actions by opposing forces, it has been used by businesses, governments, and, increasingly, natural resource managers and conservation practitioners. Scenario planning helps people move from optimum-based decision making—trying to maximize benefits and minimize losses for a single expected future—toward robust decision making, that is, trying to maximize the likelihood of some net positive outcome across a range of plausible futures.

A scenario is a plausible future, or a structured description of that future. It is not a prediction or forecast, or a means of determining which future is most likely. Indeed, when uncertainty is high, multiple very different outcomes are often equally probable. The IPCC, for instance, created a suite of scenarios that covered a range of trajectories for demographic, social, economic, technological, and environmental development,

## BOX 16.1 SCENARIO PLANNING IN PRACTICE

Peterson and others (2003b) identify six steps for scenario planning, which we illustrate here with an example from the Northern Highlands Lake District in Wisconsin (Peterson et al. 2003a).

- *Identification of focal issue.* Clearly define the focus of the exercise—particular places, species, ecosystem services, and so on. What to include in scenarios depends on what you care about.
  *Northern Highlands Lake District example*: Over the course of a year, ecologists and managers discussed the past, present, and possible future of the district, and decided to focus on water quality and fish populations.
- *Assessment of system status and function.* List internal and external sociopolitical, economic, or ecological forces that influence system dynamics. Highlight those that are important, and uncontrollable, and for which the future contains significant uncertainty.
  *Northern Highlands Lake District example*: Participants analyzed and discussed social and ecological forces within the Lake District and created a conceptual model (fig. 17.4) of actors, ecosystem components, linkages, and key external driving forces.

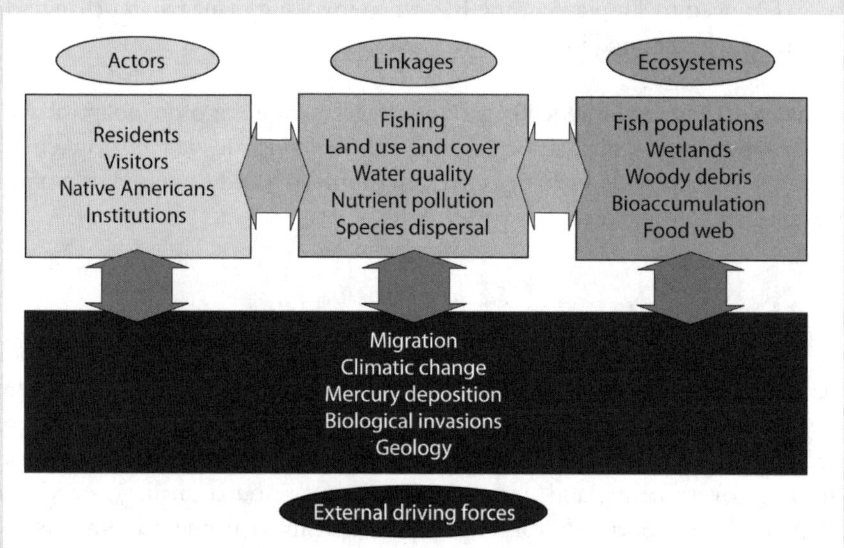


FIGURE 16.4 Conceptual model of actors, ecosystem components, linkages, and key external driving forces from the North Highlands Lake District scenario planning process. After Peterson et al. 2003a.

- *Identify a range of plausible alternative futures.* Lay out different ways the key uncertainties might unfold, and explore what the resulting futures would look like. Select a subset of the key uncertainties to help define the range of alternatives.
  *Northern Highlands Lake District example*: The group chose to focus on human migration patterns and ecological vulnerability. For instance, in one scenario warmer weather leads to population loss due to fewer winter sports opportunities and more disease, while in another the population increases because the milder climate makes for easier year-round living.
- *Create scenarios.* Turn the key alternatives from the above step into dynamic stories. Each narrative contains plausible forces and events, building seamlessly from the known past and present into the future. Each scenario describes what happens to the focal variables, and has a name that makes it memorable and accessible.
  *Northern Highlands Lake District example*: The group created three scenarios: (1) Walleye Commons, in which climate change and a globalized economy lead to decreased regional tourism and employment and many people move back to urban centers; (2) Northwoods.com, in which community development plans bring in businesses and population size and wealth soars, leading first to environmental degradation and then to successful regulation; and (3) Lake Mosaic, in which baby boomers buy vacation or retirement homes in the region and form lake associations that determine what activities are allowed on individual lakes.
- *Test scenarios.* Testing can be done by comparing scenarios with stakeholder behavior or knowledge, quantitative models, expert opinion, or other real-world information.
  *Northern Highlands Lake District example*: Project participants held an interactive workshop involving a broader array of stakeholders. This led to a final series of four full narrative scenarios to reflect what stakeholders felt were more realistic social responses to anticipated changes.
- *Use scenarios to screen policies, plans, or other actions.* Test, analyze, and create policies or action plans based on each of the scenarios. Are there common elements that work well in all scenarios? These can become core strategies. How do existing policy or management plans fare under each scenario? What monitoring plans might help managers to see which scenario seems to be unfolding?
  *Northern Highlands Lake District example*: Stay tuned. . . .

each of which leads to different levels of greenhouse gas emission, different timing of greenhouse gas stabilization, and a different rate and severity of climate change. Scenarios may be built using a formal approach that emphasizes quantifiable information and numerical analysis, or using a more intuitive, qualitative approach that makes greater use of stakeholder knowledge and input.

The goal of a scenario-planning exercise, which may range from purely exploratory to tightly focused on one particular decision, strongly influences both the way in which scenarios are developed and scenario content. Global scenarios, such as those in the Millennium Ecosystem Assessment, tend to be expert-driven, exploratory, and focused on scientific quantitative rigor. Often, an exploratory scenario process helps set the stage for a more focused approach, helping participants deepen their understanding of the system in question, stimulating creative thinking, and increasing capacity for making decisions in the face of uncertainty. Local-scale scenarios tend to be more focused on communication and consensus-building, and have a shorter time horizon than global scenarios. In some situations, scenarios can be combined across scales. Royal Dutch Shell, a pioneer in scenario planning, often developed global scenarios at the corporate level that were then used to develop a separate set of scenarios for use by the Shell operating companies in different countries. Each operating company would develop scenarios addressing the strategic issues and unknowns relevant to their operations.

There are a number of common traps people fall into during scenario planning. These include overestimating the ability of humans or institutions to control the future; blindness of participants to their own assumptions; relying more heavily on "expert" opinion than on "laypeople's" opinions (in highly uncertain situations, expert opinion may be even less reliable than that of nonexperts); overweighting present conditions; and becoming too wedded to the particular details of each scenario rather than continuing to focus on the big picture. Keeping these traps in mind may help avoid them.

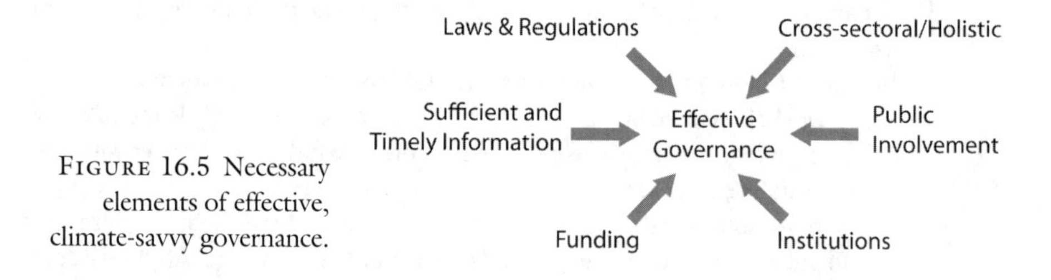


FIGURE 16.5 Necessary elements of effective, climate-savvy governance.

## Final Thoughts

Getting to climate-savvy governance and management requires the same type of holistic thinking that is needed for developing good adaptation strategies (fig. 16.5), namely thoughtful, proactive approaches that bring together stakeholders and institutions to innovate and implement flexible yet effective laws and regulations, either through adjusting existing ones or creating new ones. It's a tall order, and there is no simple recipe. Still, the benefits of a strong adaptive governance system extend beyond addressing climate change, making societies more robust to all manner of change.

## Afterword

# *Creative Thinking in Conservation and Management*

You can't wait for inspiration. You have to go after it with a club.
—*Jack London*

Climate change is a problem without boundaries. To create conservation and management solutions that work under these new conditions we cannot redeploy existing approaches without alteration and expect success. Instead, we need to think like climate change—broadly across the landscape, further into the future, and without regard to sector-based approaches. We need to blow the sides off the box and look out to the horizon, appreciating both the magnitude of the challenge and the range of options it presents. When it comes to addressing complex problems like climate change, creativity and the ability to integrate examples from multiple arenas are our best assets. As innovators of the new conservation and management path, we need to put on our thinking caps, use both sides of our brains, step out of imagined boxes (fig. A.1), and leave dogma at the door.

In this chapter, we consciously abandon the academic approach and use the musings of Yogi Berra, baseball player and philosopher, as an inspiration for thinking about conservation and resource management differently. After all, one element of coming up with new ideas and paradigms is listening to voices beyond the usual. Just as looking at the world upside-down can sometimes help you see it more clearly, so can fresh eyes and fresh voices from disparate places or disciplines help you see your own problems (and possible solutions) more clearly. Berra's famously oddball pearls of wisdom may just give us the fresh eyes we need.

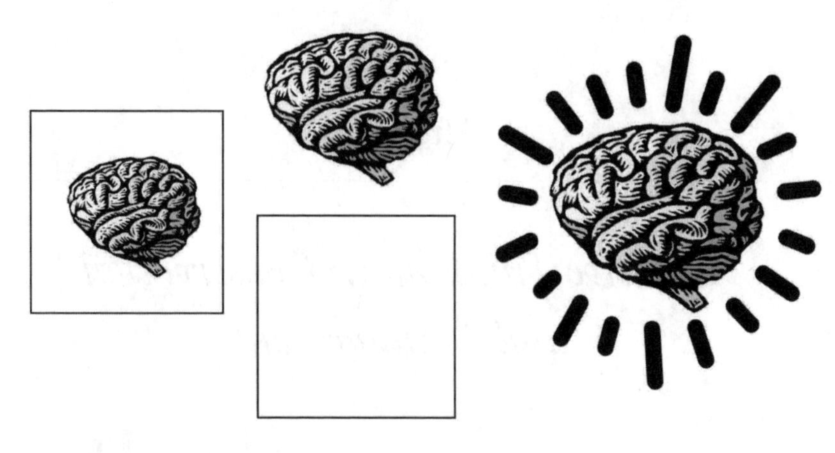


FIGURE A.1 The point is not to think outside the box, but to ignore the existence of boxes altogether.

## The Future Ain't What It Used to Be

Some aspects of the world—the basic laws of physics used for climate models, for instance—will not be affected by climate change. Many elements of ecosystems and communities that we have considered fixed, however, *will* change, sometimes radically. At some level, climate change merely emphasizes that nature is not static, a reality that has long been recognized by thoughtful individuals but was not broadly integrated into conservation and management. Global warming is hitting us over the head with the message that we need to plan for change. Rather than bracing for it, perhaps we should think more like surfers—stay in touch with the ever-changing wave of ecosystems, social systems, and climate systems and adjust our balance in response to the continuous input we receive. We will certainly wipe out, but we can get back on the surfboard and try again. Over time, we will feel comfortable surfing bigger and bigger waves, and will ride them longer and more smoothly.

## When You Come to a Fork in the Road, Take It

No one can tell us exactly what climate change will do. No model, no paleontologist, no clairvoyant knows for certain what will happen at the scales needed by managers. Then again, no one can say for certain what economic or political systems will do, or what technological innovations will arise that fundamentally change how we work (remember life before the Internet?). This is what makes adaptive governance approaches such as scenario planning so important. What short-term actions leave us with the best set of options in the medium and long term over a range of plausible futures? What information about how the future is unfolding will help us avoid pitfalls or let us know when it is time to change course? Simply waiting for the future to happen—not taking the fork in the road—both puts us at risk of failure and means we never experience the thrill of surfing.

## You Can Observe a Lot by Just Watching

Professionals in any given field tend to adopt a frame of reference or perspective learned in their formal training and work experience. We rely on these frames to simplify our life and work. They make it easier to communicate with colleagues by providing a common language, and allow us to create a standard of practice based on the experiences of our professional community. The flip side of these benefits is that we may have blind faith that what we have been doing will continue to work despite the changing conditions around us. It is vitally important to step outside of our frameworks every now and then to take stock of the bigger reality. Like a prairie dog popping its head up out of the hole, we need to see what is going on outside of our burrow. Are proposed activities in other sectors likely to increase the vulnerability of natural resources to climate change, as with the water-heavy solar power plants discussed in chapter 15? Can we develop new partnerships that allow us to increase the effectiveness of protected areas or increase landscape connectivity, as discussed in chapters 8 and 10? We may learn useful lessons by interacting across fields, biomes, or other boundaries, including how to reframe our thinking altogether if that seems like the most beneficial path forward. Cross-pollination is more important than ever for developing successful new conservation and management strategies.

## It's Déjà Vu All Over Again

Although we are urging you to brainstorm new, creative approaches to conservation and management, sometimes the answer may lie in resurrecting or reconditioning old ideas. For instance, freshwater has long been a cause of conflict for human societies. As discussed in chapter 2, freshwater is becoming even less available in some areas due to drought, loss of snowpack and glaciers, and saltwater inundation due to sea-level rise. This has inspired people to be creative not only about conserving water but also about using brackish water in place of fresh, say by thinking about more salt-resistant crops and forests. For example, cypress populations naturally have some individuals that are more salt-tolerant than others, and some coastal restoration projects are now attempting to breed more salt-tolerant variants that can withstand increasing salinity from sea-level rise. There have been significant efforts to create salt-tolerant cultivars through biotechnology, but traditional agricultural approaches have been more successful than modern technology in this regard. Indeed, some experts think salt tolerance is too complex a trait to bioengineer (Rozema and Flowers 2009). Much as we should not wait for a new technology to start reducing greenhouse gas emissions, we should take another look at the tools we already have and think about how to use them in creative new ways to deal with the challenges at hand.

## We Made Too Many Wrong Mistakes

There is a lot of rhetoric about finding opportunity in crisis, and this is not a bad attitude to adopt with climate change. The alternative is to become paralyzed by the magnitude of the problem. One hopeful aspect of climate-savvy conservation is that we can treat it as a "do over" button. As we rethink the challenges ahead and potential solutions, we can take the opportunity to correct past errors in how we have designed or carried out conservation plans. Sometimes it is difficult to accept that a plan was implemented incorrectly, lacked the right stakeholder engagement, or is actually damaging social-ecological conditions. Changing hard-fought conservation paths once they are in place can cause worry about implications for future battles. Recognizing past errors as barriers to successful climate change adaptation gives us the opportunity to explore the challenge in a whole new way and to try to come up with a better solution for both the present and the longer-term climate-informed set of conservation challenges. Climate adaptation planning allows for the formation of new alliances and aims to achieve more holistic and parsimonious solutions. A new way of thinking can lead to a new way of doing, and a whole new path forward.

## If You Don't Know Where You're Going, You Might Not Get There

When we feel unable to make decisions without more analysis, more data, or more opinions, we have achieved "analysis paralysis." We must accept that we will never have all the information we would like about what the future holds—it is the future, after all. Acting based on the best available science does not mean waiting to act until we understand everything perfectly. It means acting based on the best science that is available now. For climate change, the best available science says that there is a lot of uncertainty. Waiting too long to act can close off options and force us into conservation culs-de-sac from which it may be difficult to emerge as time (and climate change) marches on. We may do better to take a "fail early, fail often, learn quickly" approach such as the active adaptive management discussed in chapter 16. If we try and fail at least we learn something, while waiting means we may be wrong anyway and have missed other opportunities.

Fortunately, there are means for making educated guesses about the range of possibilities for the future while starting to brainstorm ideas about what we do about those possible futures. We may even discover that some actions make sense regardless of which future comes to pass, so more detailed information would not affect our decision anyway.

The conservation community likes to "strategically plan." There is nothing we need more than some good strategic planning to deal with climate change, except of course some good strategic action to deal with the problem and keep us out of the culs-de-sac. We need to make choices actively before bad choices are made for us by our lack of action.

## It's Not Too Far, It Just Seems Like It Is

Although we don't know exactly what the future will hold, we often have a pretty good idea of the general direction. On the greenhouse gas emissions front, we know that without action emissions will continue increasing and we know that we would prefer they decrease. Finding the single best action to solve the emissions problem is daunting, to say the least, but Pacala and Socolow (2004) demonstrate that we can combine multiple achievable solution wedges into a package that gives us the reduction we need (fig. A.2). These wedges include such actions as energy conservation, energy efficiency, or increasing use of renewable energy sources. In this framework, there is not one silver bullet solution and there is not one actor making all the changes. Rather, a broad array of sectors and actors come together to make the solution happen. Climate change is a big problem that needs a big solution, but that big solution will likely be built of many smaller solutions.

The same approach can be applied to adaptation. Most problems do not have a single answer that can be implemented by a single actor to achieve success. In the case of sea-level rise, the climate change issue that has garnered the most attention by planners,

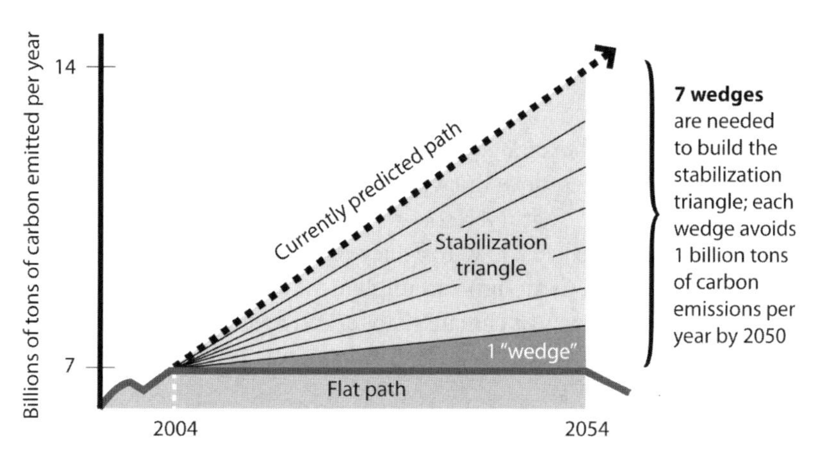


FIGURE A.2 Wedge approach to emissions stabilization. Each wedge represents an activity that reduces greenhouse gas emissions, and would be scaled up gradually such that within fifty years it would account for a 1 GtC/year reduction in carbon emissions. Pacala and Socolow (2004) describe fifteen possible wedges ranging from more efficient buildings to substituting natural gas for coal to capturing and storing carbon emitted by power plants.

there are a host of adaptation options. The list of possible wedges for adapting to sea-level rise might include:

1. Nature-based solutions
   Protect places that have no barriers to the inland movement of coastal ecosystems as sea level rises
   Protect natural features such as sediment input or coastal vegetation that help coastal accretion keep pace with sea-level rise
   Use living shorelines such as coral or oyster reefs, sea grass beds, or marshes to reduce erosion and enhance accretion rather than relying on seawalls or other shoreline hardening
   Prepare habitat behind the present shoreline to support coastal ecosystem retreat
2. Engineered solutions
   Redesign or adjust public services like water and sewage treatment for continued function and minimal environmental damage as sea levels rise
   Redirect dredging material to build up islands and coastlines to prevent them from becoming inundated
   Create structures that enhance deposition, such as artificial reefs or boulder fields
3. Policy solutions
   Reduce water withdrawals from coastal aquifers to reduce the rate of land subsidence
   Restore natural levels of sediment delivery to shorelines, for instance by removing dams, breaching levees, or prohibiting channelization
   Prohibit new construction or remove existing infrastructure in vulnerable locations, and provide sufficient setback to protect people, infrastructure, and ecosystems

None of these approaches alone will solve the problems posed by sea-level rise, but together they could greatly reduce our vulnerability to it. We need to bring together players and wisdom from the natural resource, transportation, building, public service, and science sectors, to name just a few. To achieve an outcome that we all see as successful, we need to collaboratively design and implement a host of strategies for adapting to sea-level rise, to other effects of climate change, and to stressors that are occurring in tandem with climate change.

## If the World Was Perfect, It Wouldn't Be

There is an old adage that says if you give someone a fish they eat for a day, but if you teach them to fish they can feed themselves for life. The same thinking can be applied to adaptation. Simply giving managers and practitioners a menu of adaptation options from which to choose is not a self-sustaining approach when confronted with fluid and long-term problems like climate change. Implementers may be less committed to ideas they did not develop themselves, and may have difficulty tailoring approaches to their individual social, political, or ecological circumstances. Implementation is less likely to evolve and adapt to changes in climate, economy, or other circumstances if managers and practitioners have less understanding of the thinking behind the strategies. By creating and supporting networks to share information, tools, and training, practitioners can expand their frames of reference and start innovating ways to adapt management, policies, and planning to climate change.

Give practitioners an adaptation strategy and they solve the problem for one day; teach them how to innovate adaptation strategies and they do it for a lifetime.

## It Ain't Over 'til It's Over

Creative thinking is in short supply only if we let it be. With a challenge like climate change we need to make sure the pipeline of new ideas is flowing fast and furious. Just as climate change was not created in one or two generations, it will not be resolved in our lifetimes. We need to foster good ideas for dealing with it now, and we need to build a good foundation for how it can be dealt with in the future. There are twists and turns ahead, and we need clever and ambitious plans to address what we know for certain and what we know we do not know for certain. Even more important, we need action as well as plans. We need to start implementing ideas, monitoring their efficacy, sharing the lessons, and building newer creative approaches based on what we learn. We cannot rest on our laurels, nor can we bury our heads in the sand. We have got to make conservation and resource management climate-savvy. We need to adapt conservation and resource management to climate change.

# REFERENCES

Acosta-Michlik, L. 2004. *Intervulnerability Assessment: An Innovative Framework to Assess Vulnerability to Interacting Impacts of Climate Change and Globalisation.* A project of the Advanced Institute on Vulnerability to Global Environmental Change. Washington, DC: START.

Adamcik, R. S., E. S. Bellantoni, D. C. DeLong Jr., J. H. Schomaker, D. B. Hamilton, M. K. Lauban, and R. L. Schroeder. 2004. *Writing Refuge Management Goals and Objectives: A Handbook*. Washington, DC: U.S. Fish and Wildlife Service.

A'mar, Z. T., A .E. Punt, and M. W. Dorn. 2009. The evaluation of two management strategies for the Gulf of Alaska walleye pollock fishery under climate change. *ICES Journal of Marine Science* 66: 1614–32.

Amstrup, S. C., B. G. Marcot, and D. C. Douglas. 2007. Forecasting the range-wide status of polar bears at selected times in the 21st century, *USGS Science Strategy to Support U.S. Fish and Wildlife Service Polar Bear Listing Decision*. Administrative Report. Washington, DC: U.S. Department of the Interior, U.S. Geological Survey.

Anisimov, O. A., D. G. Vaughan, T. V. Callaghan, C. Frugal, H. Marchant, T. D. Prowse, H. Vilhjálmsson, and J. E. Walsh. 2007. Polar regions (Arctic and Antarctic), in *Climate Change 2007: Impacts, Adaptation and Vulnerability: Contribution of Working Group II to the Fourth Assessment Report of the Intergovernmental Panel on Climate Change*, ed. M. L. Parry, O. F. Canziani, J. P. Palutikof, P. J. van der Linden, and C. E. Hanson, 653–85. Cambridge: Cambridge University Press.

Araki, H., B. Cooper, and M. S. Blouin. 2007. Genetic effects of captive breeding cause a rapid, cumulative fitness decline in the wild. *Science* 318: 100–103.

Association of Fish and Wildlife Agencies. 2009. *Voluntary Guidance for States to Incorporate Climate Change into State Wildlife Action Plans and Other Management Plans.* Published online at www.fishwildlife.org/pdfs/ClimateChangeGuidance%20Document_Final_reduced%20size.pdf.

Augerot, X. 2005. *Atlas of Pacific Salmon: The First Map-based Status Assessment of Salmon in the North Pacific*. Berkeley, CA: University of California Press.

Bala, G., K. Caldeira, M. Wickett, T. J. Phillips, D. B. Lobell, C. Delire, and A. Mirin. 2007. Combined climate and carbon-cycle effects of large-scale deforestation. *Proceedings of the National Academy of Sciences* 104: 6550–55.

Barnes, D. K. A., D. A. Hodgson, P. Convey, C. S. Allen, and A. Clarke. 2006. Incursion and excursion of Antarctic biota: Past, present and future. *Global Ecology and Biogeography* 15(2): 121–42.

Barnett, T. P., and D. W. Pierce. 2008. When will Lake Mead go dry? *Water Resources Research* 44: W03201, doi: 10.1029/2007WR006704.

Beale, C. M., J. J. Lennon, and A. Gimona. 2008. Opening the climate envelope reveals no macroscale associations with climate in European birds. *Proceedings of the National Academy of Sciences* 105: 14908–12.

Bell, A., N. Collins, and R. Young. 2003. *Practitioner's Guide to Incorporating Climate Change into the Environmental Impact Assessment Process.* Report prepared by ClimAdapt for the government of Nova Scotia.

Berman, M., C. Nicolson, G. Kofinas, J. Tetlichi, and S. Martin. 2004. *Adaptation and Sustainability in a Small Arctic Community: Results of an Agent-Based Simulation Model. Arctic* 57(4): 401–14.

Berteaux, D., D. Reale, A. G. McAdam, and S. Boutin. 2004. Keeping pace with fast climatic change: Can arctic life count on evolution? *Integrative and Comparative Biology* 44: 140–51.

Bettinetti, R., S. Quadroni, S. Galassi, R. Bacchetta, L. Bonardi, and G. Vailati. 2008. Is meltwater from Alpine glaciers a secondary DDT source for lakes? *Chemosphere* 73(7): 1027–31.

Blais, J. M., D. W. Schindler, D. C. G. Muir, M. Sharp, D. Donald, M. Lafrenière, E. Braekevelt, and M. M. J. Strachan. 2001. Melting glaciers: A major source of persistent organochlorines to subalpine Bow Lake in Banff National Park, Canada. *Ambio* 30(7): 410–15.

Boersma, P. D. 2008. Penguins as marine sentinels. *BioScience* 58(7): 597–607.

Bothwell, M. L., D. M. J. Sherbot, and C. M. Pollock.1994. Ecosystem response to solar ultraviolet-B radiation: Influence of trophic level interactions. *Science* 256: 97–100.

Bowyer C., D. Baldock, G. Tucker, C. Valsecchi, M. Lewis, P. Hjerp, and S. Gantioler. 2009. *Positive Planning for Onshore Wind*. London: Institute for European Environmental Policy/ Royal Society for the Protection of Birds.

Bradley, B. A., D. S. Wilcove, and M. Oppenheimer. 2009. Climate change and plant invasions: Restoration opportunities ahead? *Global Change Biology* 15: 1511–21.

Burns, C. E., K. M. Johnston, and O. J. Schmitz. 2003. Global climate change and mammalian species diversity in U.S. national parks. *Proceedings of the National Academy of Sciences* 100(20): 11474–77.

Burton, I., J. B. Smith, and S. Lenhart. 1998. Adaptation to climate change: Theory and assessment. In *Handbook on Methods for Climate Change Impact Assessment and Adaptation Strategies*, ed. J. F. Feenstra, I. Burton, and J. B. Smith, 93–116. Nairobi: UNEP.

Byers, J. E., 2002. Impact of non-indigenous species on natives enhanced by anthropogenic alteration of selection regimes. *Oikos* 97: 449–58.

Byers, J. E., and J. Pringle. 2006. Going against the flow: Retention, range limits and invasions in advective environments. *Marine Ecology Progress Series* 313: 27–41.

California Department of Boating and Waterways and State Coastal Conservancy. 2002. *California Beach Restoration Study*. Sacramento, CA. Accessed March 10, 2009, at www.dbw.ca .gov/Environmental/BeachReport.aspx.

Carter, T. R., M. L. Parry, H. Harasawa, and S. Nishioka, eds. 1994. *IPCC Technical Guidelines for Assessing Climate Change Impacts and Adaptation.* London: Department of Geography, University College.

Castellón, T. D., and K. E. Sieving. 2006. An experimental test of matrix permeability and corridor use by an endemic understory bird. *Conservation Biology* 20: 135–45.

CCAMLR (Commission for the Conservation of Antarctic Marine Living Resources) . 2009. Schedule of conservation measures in force 2009/10. (Including Resolution 30-XXVIII: Climate Change.) Accessed online at www.ccamlr.org/pu/e/e_pubs/cm/09-10/all.pdf.

Chan, F., J. A. Barth, J. Lubchenco, A. Kirincich, H. Weeks, W. T. Peterson, and B. A. Menge. 2008. Emergence of anoxia in the California current large marine ecosystem. *Science* 319(5865): 920.

Charles, S., F. Subtil, J. Kielbassa, and D. Pont. 2008. An individual-based model to describe a bullhead population dynamics including temperature variations. *Ecological Modelling* 215: 377–92.

Choi, Y. D., V. M. Temperton, E. B. Allen, A. P. Grootjans, M. Halassy, R. J. Hobbs, M. A. Naeth, and K. Torok. 2008. Ecological restoration for future sustainability in a changing environment. *Ecoscience* 15: 53–64.

Christensen, N. S., A. W. Wood, N. Voisin, D. P. Lettenmaier, and R. N. Palmer. 2004. The effects of climate change on the hydrology and water resources of the Colorado River Basin. *Climatic Change* 62: 337–63.

Clark, J. A., and E. Harvey. 2002. Assessing multi-species recovery plans under the Endangered Species Act. *Ecological Applications* 12: 655–62.

Coastal Services Center. 2004. Nonpoint Source Pollution and Erosion Comparison Tool. Accessed online at http://csc.noaa.gov/digitalcoast//tools/nspect/index.html.

Code of Federal Regulations. 2005. Title 40: Protection of the Environment.

Coenen, D., I. Porzecanski, and T. L. Crisman. 2008. Future directions in conservation and development: Incorporating the reality of climate change. *Biodiversity* 9: 106–13.

Conner, W. H., and W. Inabinette. 2005. Identification of salt tolerant baldcypress (*Taxodium distichum* (L.) Rich) for planting in coastal areas. *New Forests* 29: 305–12.

Cook, K. H., and E. K. Vizy. 2008. Effects of 21st century climate change on the Amazon rainforest. *Journal of Climate* 21: 542–60.

Coulson, S. J., I. D. Hodkinson, N. R. Webb, K. Mikkola, J. A. Harrison, and D. E. Pedgley. 2002. Aerial colonization of high Arctic islands by invertebrates: the diamondback moth *Plutella xylostella* (Lepidoptera: Yponomeutidae) as a potential indicator species. *Diversity and Distributions* 8: 327–34.

Cox, P. M., R. A. Betts, M. Collins, P. P. Harris, C. Huntingford, and C. D. Jones. 2004. Amazonian forest dieback under climate-carbon cycle projections for the 21st century. *Theoretical and Applied Climatology* 78: 137–56.

Cury, P. M., Y.-J. Shin, B. Planque, J. M. Durant, J.-M. Fromentin, S. Kramer-Schadt, N. C. Stenseth, M. Travers, and V. Grimm. 2008. Ecosystem oceanography for global change in fisheries. *Trends in Ecology and Evolution* 23(6): 338–46.

Davidson, I., and C. Simkanin. 2008. Skeptical of assisted colonization. *Science* 322: 1048–49.

Davies, R. W. D., S. J. Cripps, A. Nickson, and G. Porter. 2009. Defining and estimating global marine fisheries bycatch. *Marine Policy* 33: 661–72.

Davis, M. B., and R. G. Shaw. 2001. Range shifts and adaptive responses to quaternary climate change. *Science* 292: 673–79.

de Bruin, K., R. B. Dellink, A. Ruijs, L. Bolwidt, A. van Buuren, J. Graveland, R. S. de Groot, et al. 2009. Adapting to climate change in the Netherlands: An inventory of climate adaptation options and ranking of alternatives. *Climatic Change* 95: 23–45.

de Bruin, K. C., R. B. Dellink, and R. S. J. Tol. 2007. AD-DICE: An implementation of adaptation in the DICE model. *Working Papers FNU-126*, Hamburg, Germany: Research Unit Sustainability and Global Change, Hamburg University.

Deering, A., and J. P. Thornton. 1998. *Solar Technology and the Insurance Industry: Issues and Applications*. NREL/MP 520-25866, Golden, CO: National Renewable Energy Laboratory.

DEFRA (Department for Environment, Food and Rural Affairs). 2006. *Shoreline Management Plan Guidance*.

Diaz, R. J., and R. Rosenberg. 2008. Spreading dead zones and consequences for marine ecosystems. *Science* 321: 926–29.

Dickinson, T. 2007. *The Compendium of Adaptation Models for Climate Change*, first edition. Ottawa: Adaptation and Impacts Research Division, Environment Canada.

Diederich, S., G. Nehls, J. E. E. van Beusekom, and K. Reise. 2005. Introduced Pacific oysters (*Crassostrea gigas*) in the northern Wadden Sea: Invasion accelerated by warm summers? *Helgoland Marine Research* 59(2): 97–106.

Dietz, T., E. Ostrom, and P. C. Stern. 2003. The struggle to govern the commons. *Science* 302: 1907–12.

Di Tomaso, J. M. 1998. Impact, biology, and ecology of saltcedar (Tamarix spp.) in the southwestern United States. *Weed Technology* 12: 326–36.

Duvall, S. E., and M. G. Barron. 2000. A screening level probabilistic risk assessment of mercury in Florida Everglades food webs. *Ecotoxicology and Environmental Safety* 47(3): 298–305.

Erwin, R. M., J. Miller, and J. G. Reese. 2007. Poplar Island Environmental Restoration Project: Challenges in waterbird restoration on an island in Chesapeake Bay. *Ecological Restoration* 25: 256–62.

FAO (Food and Agriculture Organization). 2005. *Review of the State of World Marine Fishery Resources*. FAO Fisheries Technical Paper 457. Accessed online at www.fao.org/docrep/009/y5852e/Y5852E00.htm#TOC.

Feely, R. A., C. L. Sabine, J. M. Hernandez-Ayon, D. Ianson, and B. Hales. 2008. Evidence for upwelling of corrosive "acidified" water onto the Continental Shelf. *Science* 320: 1490–92.

Field, J. C., R. C. Francis, and K. Aydin. 2006. Top-down modeling and bottom-up dynamics: Linking a fisheries-based ecosystem model with climate hypotheses in the Northern California Current. *Progress in Oceanography* 68: 238–70.

Fisher, N. S., and C. F. Wurster. 1973. Individual and combined effects of temperature and polychlorinated biphenyls on the growth of three species of phytoplankton. *Environmental Pollution* 5(3): 205–12.

Foden, W., G. Mace, J.-C. Vié, A. Angulo, S. Butchart, L. DeVantier, H. Dublin, A. Gutsche, S. Stuart, and E. Turak. 2008. Species susceptibility to climate change impacts. In *2008 Review of the IUCN Red List of Threatened Species*, ed. J.-C. Vié, C. Hilton-Taylor, and S. N. Stuart. Gland, Switzerland: IUCN.

Folke, C., T. Hahn, P. Olsson, and J. Norberg. 2005. Adaptive governance of social-ecological systems. *Annual Review of Environmental Resources* 30: 441–73.

Ford, S. E., and R. Smolowitz. 2007. Infection dynamics of an oyster parasite in its newly expanded range. *Marine Biology* 151: 119–33.

Fowler, H. J., and D. R. Archer. 2006. Conflicting signals of climatic change in the upper Indus Basin. *Journal of Climate* 19: 4276–93.

Galloway, J. N., A. R. Townsend, J. W. Erisman, M. Bekunda, Z. Cai, J. R. Freney, L. A. Martinelli, S. P. Seitzinger, and M. A. Sutton. 2008. Transformation of the nitrogen cycle: Recent trends, questions, and potential solutions. *Science* 320(5878): 889–92.

Game, E. T., E. McDonald-Madden, M. L. Puotinen, and H. P. Possingham. 2008. Should we protect the weak or the strong? Risk, resilience, and the selection of marine protected areas. *Conservation Biology* 22(6): 1619–29.

Gerlach, J. 2008. Climate change and identification of terrestrial protected areas in the Seychelles Islands. *Biodiversity* 9: 24–29.

Gibbard, S., K. Caldeira, G. Bala, T. J. Phillips, and M. Wickett. 2005. Climate effects of global land cover change. *Geophysical Research Letters* 32(23): L23705.

Graham, R. W., and E. C. Grimm. 1990. Effects of global climate change on the patterns of terrestrial biological communities. *Trends in Ecology and Evolution* 5: 289–92.

Gray, L. 2009. U.K. in danger of becoming "ecological desert." *Telegraph* 9 March. Accessed online at www.telegraph.co.uk/earth/earthnews/4948093/UK-in-danger-of-becoming-ecological-desert.html.

Gresh, T. U., J. Lichatowich, and P. Schoonmaker. 2000. An estimation of historic and current levels of salmon production in the northeast Pacific ecosystem: Evidence of a nutrient deficit in the freshwater system of the Pacific Northwest. *Fisheries* 25: 15–21.

Grier, J. W. 1982. Ban of DDT and subsequent recovery of reproduction in bald eagles. *Science* 218: 1232–35.

Groom, M. J., G. K. Meffe, and C. R. Carroll, eds. 2006. *Principles of Conservation Biology*, third edition. Sunderland, MA: Sinauer Associates.

Grothmann, T., and A. Patt. 2005. Adaptive capacity and human cognition: The process of individual adaptation to climate change. *Global Environmental Change* 15: 199–213.

Gutierrez, A. P., L. Ponti, T. d'Oultremont, and C. K. Ellis. 2008. Climate change effects on poikilotherm tritrophic interactions. *Climatic Change* 87(Suppl 1): S167–S192.

Hamm, L., M. Capobianco, H. H. Dette, A. Lechuga, R. Spanhoff, and M. J. F. Stive. 2002. A summary of European experience with shore nourishment. *Coastal Engineering* 47: 237–64.

Hanson, H., A. Brampton, M. Capobianco, H. H. Dette, L. Hamm, C. Laustrup, A. Lechuga, and R. Spanhoff. 2002. Beach nourishment projects, practices and objectives: A European overview. *Coastal Engineering* 47: 81–111.

Hayhoe, K., C. P. Wake, B. Anderson, X.-Z. Liang, E. Maurer, J. Zhu, J. Bradbury, A. DeGaetano, A. Hertel, and D. Wuebbles. 2008. Regional climate change projections for the northeast U.S.A. *Mitigation and Adaptation Strategies for Global Change* 13(5–6): 425–36.

Heller, N. E., N. J. Sanders, J. W. Shors, and D. M. Gordon. 2008. Rainfall facilitates the spread, and time alters the impact, of the invasive Argentine ant. *Oecologia* 155: 385–95.

Heller, N. E., and E. S. Zavaleta. 2009. Biodiversity management in the face of climate change: A review of 22 years of recommendations. *Biological Conservation* 142: 14–32.

Hoegh-Guldberg, O., L. Hughes, S. McIntyre, D. B. Lindenmayer, C. Parmesan, H. P. Possingham, and C. D. Thomas. 2008. Assisted colonization and rapid climate change. *Science* 321: 345–46.

Hohenlohe, P. 2004. Limits to gene flow in marine animals with planktonic larvae: models of Littorina species around Point Conception, California. *Biological Journal of the Linnean Society* 82: 169–87.

Hollowed, A. B., N. A. Bond, T. K. Wilderbuer, W. T. Stockhausen, Z. T. A'mar, R. J. Beamish, J. E. Overland, and M. J. Schirripa. 2009. A framework for modelling fish and shellfish responses to future climate change. *ICES Journal of Marine Science* 66: 1584–94.

Hood, G. A., and S. E. Bayley. 2008. Beaver (*Castor canadensis*) mitigate the effects of climate on the area of open water in boreal wetlands in western Canada. *Biological Conservation* 141: 556–67.

Hughes, T. P., M. J. Rodrigues, D. R. Bellwood, D. Ceccarelli, O. Hoegh-Guldberg, L. McCook, N. Moltschaniwskyj, M. S. Pratchett, R. S. Steneck, and B. Willis. 2007. Phase shifts, herbivory and the resilience of coral reefs to climate change. *Current Biology* 17: 360–65.

Hulme, P. E. 2009. Trade, transport and trouble: managing invasive species pathways in an era of globalization. *Journal of Applied Ecology* 46: 10–18.

Huntingford, C., R. A. Fisher, L. Mercado, B. B. Booth, S. Sitch, P. P. Harris, P. M. Cox, et al. 2008. Towards quantifying uncertainty in predictions of Amazon "dieback." *Philosophical Transactions of the Royal Society of London B: Biological Sciences* 363: 1857–64.

Idaho Sage-grouse Advisory Committee. 2006. *Conservation Plan for the Greater Sage-grouse in Idaho*. Accessed online at http://fishandgame.idaho.gov/cms/hunt/grouse/conserve_plan/.

Integrated Resource Planning Committee. 1993. *Public Participation Guidelines for Land and Resource Management Planning*. Victoria, BC: The Land Use Coordination Office, Province of British Columbia.

IPCC. 2001. *Synthesis Report of the Third Assessment Report*. Cambridge: Cambridge University Press.

IPCC. 2007a. *Climate Change 2007: The Physical Science Basis. Contribution of Working Group I to the Fourth Assessment Report of the Intergovernmental Panel on Climate Change*. Ed. S. Solomon, D. Qin, M. Manning, Z. Chen, M. Marquis, K. B. Averyt, M. Tignor, and H. L. Miller. Cambridge: Cambridge University Press.

IPCC. 2007b. *Climate Change 2007: Impacts, Adaptation and Vulnerability. Contribution of Working Group II to the Fourth Assessment Report of the Intergovernmental Panel on Climate Change*. Ed. M. L. Parry, O. F. Canziani, J. P. Palutikof, P. J. van der Linden, and C. E. Hanson. Cambridge: Cambridge University Press.

Jackson, J. B. C., M. X. Kirby, W. H. Berger, K. A. Bjorndal, L. W. Botsford, B. J. Bourque, R. H. Bradbury et al. 2001. Historical overfishing and the recent collapse of coastal ecosystems. *Science* 293: 629–37.

Jackson, S. T., and R. J. Hobbs. 2009. Ecological restoration in the light of ecological history. *Science* 325: 567–69.

Jefferies, R. L., and R. H. Drent. 2006. Arctic geese, migratory connectivity and agricultural change: Calling the sorcerer's apprentice to order. *Ardea* 94 (3): 537–54.

Jessen, S., and S. Patton. 2008. Protecting marine biodiversity in Canada: Adaptation options in the face of climate change. *Biodiversity* 9: 47–58.

Jones, C. G., and J. L. Gutiérrez. 2007. On the purpose, meaning, and usage of the physical ecosystem engineering concept. In *Ecosystem Engineers (Theoretical Ecology Series)*, ed. K. Cuddington, J. E. Byers, A. Hastings, and W. G. Wilson, 3–24. Burlington, MA: Academic Press.

Julius, S. H., J. M. West, J. S. Baron, B. Griffith, L. A. Joyce, B. D. Keller, M. A. Palmer, C. H. Peterson, and J. M. Scott. 2008. Annex A: Case Studies. In *Preliminary Review of Adaptation Options for Climate Sensitive Ecosystems and Resources: A Report by the U.S. Climate Change Science Program and the Subcommittee on Global Change Research*, ed. S.H. Julius and J. M. West, A-1–A-170. Washington, DC: U.S. Environmental Protection Agency.

Keeling, C. D., and T. P. Whorf. 2004. Atmospheric $CO_2$ records from sites in the SIO air sampling network. In *Trends: A Compendium of Data on Global Change*. Oak Ridge,TN: Carbon Dioxide Information Analysis Center, Oak Ridge National Laboratory, U.S. Department of Energy.

Kehrwald, N. M., L. G. Thompson, Y. Tandong, E. Mosley-Thompson, U. Schotterer, V. Alfimov, J. Beer, J. Eikenberg, and M. E. Davis. 2008. Mass loss on Himalayan glacier endangers water resources. *Geophysical Research Letters* 35: L22503, doi:10.1029/2008 GL035556.

Kingdom of Bhutan. 2008. The Constitution of the Kingdom of Bhutan, Article 5(3).

Kitheka, J. U., G. S. Ongwenyi, and K. M. Mavuti. 2002. Dynamics of suspended sediment exchange and transport in a degraded mangrove creek in Kenya. *Ambio* 31: 580–87.

Knowles, N., M. D. Dettinger, and D. R. Cayan. 2006. Trends in snowfall versus rainfall in the western United States. *Journal of Climate* 19: 4545–59.

Kutz, S. J., E. P. Hoberg, J. Nagy, L. Polley, and B. Elkin. 2004. "Emerging" parasitic infections in arctic ungulates. *Integrative and Comparative Biology* 44: 109–18.

Lacan, I., K. R. Matthews, and K. V. Feldman. 2008. Interaction of an introduced predator with future effects of climate change in the recruitment dynamics of the imperiled Sierra Nevada yellow-legged frog (*Rana sierrae*). *Herpetological Conservation and Biology* 3(2): 211–23.

LaFever, D. H., R. R. Lopez, R. A. Feagin, and N. J. Silvy. 2007. Predicting the impacts of future sea-level rise on an endangered lagomorph. *Environmental Management* 40: 430–37.

Laidre, K. L., I. Stirling, L. F. Lowry, ø. Wiig, M. P. Heide-Jorgensen, and S. H. Ferguson. 2008. Quantifying the sensitivity of Arctic marine mammals to climate-induced habitat change. *Ecological Applications* 18(Supplement): S97–S125.

Lansing, J. S. 1991. *Priests and Programmers: Technologies of Power in the Engineered Landscape of Bali*. Princeton, NJ: Princeton University Press.

Lawler, J. J. 2009. Climate change adaptation strategies for resource management and conservation planning. *Annals of the New York Academy of Sciences* 1162: 79–98.

Lawton, R. O., U. S. Nair, R. A. Pielke Sr., and R. M. Welch. 2001. Climatic impact of tropical lowland deforestation on nearby montane cloud forests. *Science* 294: 584–87.

Leung, R., Y. Qian, X. Bian, W. Washington, J. Han, and J. O. Roads. 2004. Mid-century ensemble regional climate change scenarios for the western United States. *Climatic Change* 62: 75–113.

Lopez, R. R., N. J. Silvy, R. N. Wilkins, P. A. Frank, M. J. Peterson, and M. N. Peterson. 2004. Habitat-use patterns of Florida Key deer: Implications of urban development. *Journal of Wildlife Management* 68(4): 900–908.

Marshall, P. A., and L. E. Johnson. 2007. The Great Barrier Reef and climate change: Vulnerability and management implications. In *Climate Change and the Great Barrier Reef: A Vulnerability Assessment*, ed. J. E. Johnson and P. A. Marshall, 773–802. Townsville, Australia: Great Barrier Reef Marine Park Authority.

Marshall, P. A., and H. Shuttenberg. 2006. *A Reef Manager's Guide to Coral Bleaching*. Townsville, Australia: Great Barrier Reef Marine Park Authority.

Marty, J. T. 2005. Effects of cattle grazing on diversity in ephemeral wetlands. *Conservation Biology* 19(5): 1626–32.

Matthews, K. R., K. Feldman, and H. Preisler. 2009. Seasonal habitat use and site fidelity of the mountain yellow-legged frog in Sierra Nevada high-elevation lakes. Accessed June 23, 2009, at www.fs.fed.us/psw/programs/snrc/bio_diversity/aquatic_lessdistrubed_sub2/site_fidelity.shtml.

McClenachan, L., J. B. C. Jackson, and M. Hardt. 2006. Conservation implications of historic sea turtle nesting beach loss. *Frontiers in Ecology and the Environment* 4(6): 290–96.

McRae, B. H., B. G. Dickson, T. H. Keitt, and V. B. Shah. 2008. Using circuit theory to model connectivity in ecology, evolution, and conservation. *Ecology* 89(10): 2712–24.

Meehl, G. A., W. M. Washington, W. D. Collins, J. M. Arblaster, A. Hu, L. E. Buja, W. G. Strand, and H. Teng. 2005. Howvmuch more global warming and sea level rise? *Science* 307: 1769–72.

Meffe, G. K., and S. Viederman. 1995. Combining science and policy in conservation biology. *Wildlife Society Bulletin* 23(3): 327–32.

Meko, D. M., C. A. Woodhouse, C. A. Baisan, T. Knight, J. J. Lukas, M. K. Hughes, and M. W. Salzer. 2007. Medieval drought in the upper Colorado River Basin. *Geophysical Research Letters* 34: L10705, doi: 10.1029/2007GL029988.

Millar, C. I., N. L. Stephenson, and S. L. Stephens. 2007. Climate change and forests of the future: Managing on the face of uncertainty. *Ecological Applications* 17: 2145–51.

Millar, C. I., and W. B. Woolfenden. 1999. The role of climate change in interpreting historical variability. *Ecological Applications* 9(4): 1207–16.

Milly, P. C. D. 2007. Stationarity is dead. *Ground Water News & Views* 4(1): 6, 8.

Monnin, E., E. J. Steig, U. Siegenthaler, K. Kawamura, J. Schwander, B. Stauffer, T. F. Stocker, et al. 2004. Evidence for substantial accumulation rate variability in Antarctica during the Holocene, through synchronization of $CO_2$ in the Taylor Dome, Dome C, and DML ice cores. *Earth and Planetary Science Letters* 224: 45–54.

Mote, P., A. Petersen, S. Reeder, H. Shipman, and L. Whitely Binder. 2008. *Sea Level Rise in the Coastal Waters of Washington State*. Report prepared by the Climate Impacts Group, Center for Science in the Earth System, Joint Institute for the Study of the Atmosphere and Oceans, University of Washington, Seattle, WA and the Washington Department of Ecology, Lacey, WA.

Mueller, J. M., and J. J. Hellmann. 2008. An assessment of invasion risk from assisted migration. *Conservation Biology* 22: 562–67.

Naiman, R. J., R. E. Bilby, D. E. Schindler, and J. M. Helfield. 2002. Pacific salmon, nutrients and the dynamics of freshwater and riparian ecosystems. *Ecosystems* 5: 399–417.

National Marine Fisheries Service. 2008. *Conservation Plan for the Cook Inlet Beluga Whale* (Delphinapterus leucas). Juneau, AK: National Marine Fisheries Service.

National Research Council (U.S.). 1992. *Restoration of Aquatic Ecosystems: Science, Technology, and Public Policy*. Washington, DC: National Academies Press.

Neftel, A., H. Friedli, E. Moor, H. Lötscher, H. Oeschger, U. Siegenthaler, and B. Stauffer. 1994. Historical $CO_2$ record from the Siple Station ice core. In *Trends: A Compendium of Data on Global Change*. Oak Ridge, TN: Carbon Dioxide Information Analysis Center, Oak Ridge National Laboratory, U.S. Department of Energy.

Nellemann, C., S. Hain, and J. Alder (eds). 2008. *In Dead Water: Merging of Climate Change with Pollution, Over-Harvest, and Infestations in the World's Fishing Grounds*. Norway: United Nations Environment Programme, RID-Arendal.

Nelson, H. 2007. Does a crisis matter? Forest policy responses to the mountain pine beetle epidemic in British Columbia. *Canadian Journal of Agricultural Economics* 55: 459–70.

Nichols, J. D., M. C. Runge, F. A. Johnson, and B. K. Williams. 2007. Adaptive harvest management of North American waterfowl populations: A brief history and future prospects. *Journal of Ornithology* 148 (Suppl 2): S343–S349.

Nordhaus, W. D. 1992. *The "DICE" Model: Background and Structure of a Dynamic Integrated Climate-Economy Model of the Economics of Global Warming*. Cowles Foundation Discussion Papers 1009. Cowles Foundation, Yale University.

Noyes, P. D., M. K. McElwee, H. D. Miller, B. W. Clark, L. A. Van Tiem, K. C. Walcott, K. N. Erwin, and E. D. Levin. 2009. The toxicology of climate change: Environmental contaminants in a warming world. *Environmental International* 35: 971–86.

NPFMC (North Pacific Fisheries Management Council). 2009. *Fisheries Management Plan for Fish Resources of the Arctic Management Area: Public Review Draft*. Accessed online at www .fakr.noaa.gov/npfmc/current_issues/Arctic/ArcticFMP109.pdf.

Ohlemüller, R., S. Walker, and J. B. Wilson. 2006. Local vs. regional factors as determinants of the invasibility of indigenous forest fragments by alien plant species. *Oikos* 112: 493–501.

Pacala, S., and R. Socolow. 2004. Stabilization wedges: Solving the climate problem for the next 50 years with current technologies. *Science* 305: 968–72.

Page, S. E., F. Siegert, J. O. Rieley, H.-D. V. Boehm, D. Jaya, and S. Limin. 2002. The amount of carbon released from peat and forest fires in Indonesia during 1997. *Nature* 420: 61–65.

Paine, R. T. 1969. A note on trophic complexity and species diversity. *American Naturalist* 103: 91–93.

Parmesan, C. 2006. Ecological and evolutionary responses to recent climate change. *Annual Review of Ecology, Evolution and Systematics* 37: 637–69.

Patt, A., and B. Siebenhüner. 2005. Agent-based modeling and adaptation to climate change. *Vierteljahrshefte zur Wirtschaftsforschung* 74(2): 310–20.

Pauly, D., V. Christensen, J. Dalsgaard, R. Froese, and F. Torres Jr. 1998. Fishing down marine food webs. *Science* 279(5352): 860–63.

Pearsall, S. H. III, and B. Poulter. 2005. Adapting coastal lowlands to rising seas. In *Principles of Conservation Biology,* third edition, ed. M. J. Groom, G. K. Meffe, and C. R. Carroll, 366–70. Sunderland, MA: Sinauer Associates.

Pearson, R. G., W. Thuiller, M. B. Araújo, L. Brotons, E. Martinez-Meyer, C. McClean, L. Miles, P. Segurado, T. P. Dawson, and D. Lees. 2006. Model-based uncertainty in species' range prediction. *Journal of Biogeography* 33: 1704–11.

Perz, S., S. Brilhante, F. Brown, M. Caldas, S. Ikeda, E. Mendoza, C. Overdevest, V. Reis, J. F. Reyes, D. Rojas, M. Schmink, C. Souza, and R. Walker. 2008. Road building, land use and climate change: Prospects for environmental governance in the Amazon. *Philosophical Transactions of the Royal Society B: Biological Sciences* 363: 1889–95.

Peterson, G. D., T. D. Beard Jr., B. E. Beisner, E. M. Bennett, S. R. Carpenter, G. S. Cumming, C. L. Dent, and T. D. Havlicek. 2003a. Assessing future ecosystem services: A case study of the Northern Highland Lake District, Wisconsin. *Conservation Ecology* 7(3): 1. Available online at www.consecol.org/vol7/iss3/art1.

Peterson, G. D., G. S. Cummings, and S. R. Carpenter. 2003b. Scenario planning: A tool for conservation in an uncertain world. *Conservation Biology* 17(2): 358–66.

Petit, J. R., J. Jouzel, D. Raynaud, N. I. Barkov, J. M. Barnola, I. Basile, M. Bender, et al. 1999. Climate and atmospheric history of the past 420,000 years from the Vostok Ice Core, Antarctica. *Nature* 399: 429–36.

Phillips, S. J., P. Williams, G. Midgley, and A. Archer. 2008. Optimizing dispersal corridors for the Cape Proteaceae using network flow. *Ecological Applications* 18: 1200–1211.

Pittock, J., L. J. Hansen, R. Abell. 2008. Running dry: Freshwater biodiversity, protected areas and climate change. *Biodiversity* 9: 30–38.

Povilitis, A., and K. Suckling. 2010. Addressing climate change threats to endangered species in U.S. recovery plans. *Conservation Biology* 24(2): 372–76.

Preston, K. L., J. T. Rotenberry, R. A. Redak, and M. F. Allen. 2008. Habitat shifts of endangered species under altered climate conditions: Importance of biotic interactions. *Global Change Biology* 14: 2501–15.

Pretty, J. N., I. Guijt, J. Thompson, and I. Scoones. 1995. *Participatory Learning and Action: A Trainer's Guide*. IIED Training Materials Series No. 1. London: IIED.

Pyke, C. R., and J. T. Marty. 2005. Cattle grazing mediates climate change impacts on ephemeral wetlands. *Conservation Biology* 19: 1619–25.

Pyke, C. R., R. Thomas, R. D. Porter, J. J. Hellmann, J. S. Dukes, D. M. Lodge, and G. Chavarria. 2008. Current practices and future opportunities for policy on climate change and invasive species. *Conservation Biology* 22: 585–92.

Rahn, M. E., H. Doremus, and J. Diffendorfer. 2006. Species coverage in multispecies habitat conservation plans: Where's the science? *Bioscience* 56: 613–19.

Randall, D. A., R. A. Wood, S. Bony, R. Colman, T. Fichefet, J. Fyfe, V. Kattsov, et al. 2007. Climate models and their evaluation. In *Climate Change 2007: The Physical Science Basis. Contribution of Working Group I to the Fourth Assessment Report of the Intergovernmental Panel on Climate Change*, ed. S. Solomon, D. Qin, M. Manning, Z. Chen, M. Marquis, K. B. Averyt, M. Tignor, and H. L. Miller, 590–662. Cambridge: Cambridge University Press.

Ray, D. K., U. S. Nair, R. O. Lawton, R. M. Welch, and R. A. Pielke Sr. 2006. Impact of land use on Costa Rican tropical montane cloud forests: Sensitivity of orographic cloud formation to deforestation in the plains. *Journal of Geophysical Research* 111: D02108.

Rayfield, B., P. M. A. James, A. Fall, and M.-J. Fortin. 2008. Comparing static versus dynamic protected areas in the Quebec boreal forest. *Biological Conservation* 141: 438–49.

Rayfield, B., A. Moilanen, and M. Fortin. 2009. Incorporating consumer-resource spatial interactions in reserve design. *Ecological Modeling* 220: 725–33.

Regehr, E. V., N. J. Lunn, S. C. Amstrup, and I. Stirling. 2007. Effects of earlier sea ice breakup on survival and population size of polar bears in western Hudson Bay. *Journal of Wildlife Management* 71(8): 2673–83.

Remy, M. I. 1998. Se prepara Piura para El Niño. *Quehacer* 109(112): 69–71.

Rice, K. J., and N. C. Emery. 2003. Managing microevolution: Restoration in the face of global change. *Frontiers in Ecology and the Environment* 1(9): 469–78.

Richardson, D. M., J. J. Hellmann, J. S. McLachlan, D. F. Sax, M. W. Schwartz, P. Gonzalez, E. J. Brennan, et al. 2009. Multidimensional evaluation of managed relocation. *Proceedings of the National Academy of Sciences* 106(24): 9721–24.

Ripple, W. J., and R. L. Beschta. 2007. Restoring Yellowstone's aspen with wolves. *Biological Conservation* 138: 514–19.

Rosenberg, N. J., D. J. Epstein, D. Wang, L. Vail, R. Srinivasan, and J. G. Arnold. 1999. Possible impacts of global warming on the hydrology of the Ogallala Aquifer region. *Climatic Change* 42: 677–92.

Ross, L. C., P. W. Lambdon, and P. E. Hulme. 2008. Disentangling the roles of climate, propagule pressure and land use on the current and potential elevational distribution of the invasive weed *Oxalis pes-caprae* L. on Crete. *Perspectives in Plant Ecology, Evolution and Systematics* 10: 251–58.

Ross, M. S., J. J. O'Brien, R. G. Ford, K. Zhang, and A. Morkill. 2009a. Disturbance and the rising tide: The challenge of biodiversity management on low-island ecosystems. *Frontiers in Ecology and the Environment* 7 doi: 10.1890/070221.

Ross, M. S., J. J. O'Brien, and L. da Silveira Lobo Sternberg. 1994. Sea-level rise and the reduction in pine forests in the Florida Keys. *Ecological Applications* 4(1): 144–56.

Ross, P. S., C. M. Couillard, M. G. Ikonomou, S. C. Johannessen, M. Lebeuf, R. W. Macdonald, and G. Tomy. 2009b. Large and growing environmental reservoirs of deca-BDE present an emerging health risk for fish and marine mammals. *Marine Pollution Bulletin* 58: 7–10.

Rozema, J., and T. Flowers. 2009. Crops for a salinized world. *Science* 322: 1478–80.

Salazar, L. F., C. A. Nobre, and M. D. Oyama. 2007. Climate change consequences on the biome distribution in tropical South America. *Geophysical Research Letters* 34: L09708, doi: 10.1029/2007GL029695.

Sauchyn, D., and S. Kulshreshtha. 2008. The prairies. In *From Impacts to Adaptation: Canada in a Changing Climate 2007*, ed. D. S. Lemmen, F. J. Warren, J. Lacroix, and E. Bush, 275–328. Ottawa: Government of Canada.

Seastedt, T. R., R. J. Hobbs, and K. N. Suding. 2008. Management of novel ecosystems: Are novel approaches required? *Frontiers in Ecology and the Environment* 6: 547–53.

Shaffer, G., S. M. Olsen, and J. O. P. Pederson. 2009. Long-term oxygen depletion in response to carbon dioxide emissions from fossil fuels. *Nature Geoscience* 2: 10–109.

Sibold, J. S., T. T. Veblen, K. Chipko, L. Lawson, E. Mathis, and J. Scott. 2007. Influences of secondary disturbances on lodgepole pine stand development in Rocky Mountain National Park. *Ecological Applications* 17(6): 1638–55.

Sleeman, J. C., G. S. Boggs, B. C. Radford, and G. A. Kendrick. 2005. Using agent-based models to aid reef restoration: Enhancing coral cover and topographic complexity through the spatial arrangement of coral transplants. *Restoration Ecology* 13(4): 685–94.

Smith, L. C., Y. Sheng, G. M. MacDonald, and L. D. Hinzman. 2005. Disappearing arctic lakes. *Science* 308: 1429.

Sokolova, I. M., and G. Lannig. 2008. Interactive effects of metal pollution and temperature on metabolism in aquatic ectotherms: Implications of global climate change. *Climate Research* 37: 181–201.

Sorte, C. J. B., S. L. Williams, and J. T. Carlton. 2010. Marine range shifts and species introductions: Comparative spread rates and community impacts. *Global Ecology and Biogeography* 19: 303–16.

Soulé, M. E. 1985. What is conservation biology? *BioScience* 35(11): 727–34.

Soulé, M. E. 1986. *Conservation Biology: The Science of Scarcity and Diversity*. Sunderland, MA: Sinauer Associates.

Soulé, M. E., ed. 1987. *Viable Populations for Conservation*. Cambridge: Cambridge University Press.

Stachowicz, J. J., J. R. Terwin, R. B. Whitlatch, and R. W. Osman. 2002. Linking climate change and biological invasions: Ocean warming facilitates nonindigenous species invasion. *Proceedings of the National Academy of Sciences* 99: 15497–500.

Stern, N. H. 2007. *The Economics of Climate Change: The Stern Review*. Cambridge: Cambridge University Press.

Strayer, D. L. 2010. Alien species in fresh waters: Ecological effects, interactions with other stressors, and prospects for the future. *Freshwater Biology* 55(S1): 152–74.

Tan, C. S., and W. D. Reynolds. 2003. Impacts of recent climate trends on agriculture in southwestern Ontario. *Canadian Water Resources Journal* 28: 87–97.

Tano, M. L. 2006. *Developing Agile Tribal Leaders and Agile Tribal Institutions to Adaptively Manage and Mitigate the Impacts of Global Climate Change in Indian Country*. Report to International Institute for Indigenous Resource Management. Denver, CO.

Tewksbury, J. J., R. B. Huey, and C. A. Deutsch. 2008. Putting the heat on tropical animals. *Science* 320(5881): 1296–97.

Theoharides, K. A., and J. S. Dukes. 2007. Plant invasion pattern and process: Factors affecting plant invasion at four spatio-temporal stages. *New Phytologist* 176: 256–73.

Thier, A. 2001. Balancing the risks: Vector control and pesticide use in response to emerging illness. *Journal of Urban Health: Bulletin of the New York Academy of Medicine* 78(2): 372–81.

Titus, J. G. 1990. Strategies for adapting to the greenhouse effect. *Journal of the American Planning Association* 56(3): 311–23.

Titus, J. G., and J. Wang. 2008. *Maps of Lands Close to Sea Level along the Mid-Atlantic Coast*. U.S. Environmental Protection Agency. Accessed online at http://maps.risingsea.net/.

Trigoso, E. T. 2007. *Climate Change Impacts and Adaptation in Peru: The Case of Puno and Piura*. Human Development Report 2007/08.

U.N. Environmental Program. 2007. *Global Environmental Outlook GEO-4*. Valletta, Malta: Progress Press.

UNFCCC. 2008. *Compendium of Decision Tools to Evaluate Strategies for Adaptation to Climate Change*. United Nations Framework Convention on Climate Change Secretariat, Bonn, Germany.

U.S. Army Corps of Engineers. 2003. *Water Management for the Pacific Northwest Reservoir System*. Report available at www.nwd-wc.usace.army.mil/pdf/wmbroch.pdf; 2003 map update available at www.nwd-wc.usace.army.mil/report/colmap.htm.

U.S. Bureau of Land Management and California Energy Commission. 2009. *Final Staff Assessment and Draft Environmental Impact Statement and Draft California Desert Conservation Area Plan Amendment for the Ivanpah Solar Electric Generating System.*

U.S. Fish and Wildlife Service. 2008. *Draft Comprehensive Conservation Plan and Environmental Assessment. Lower Florida Keys National Wildlife Refuges*. Department of Interior, Southeast Region, Atlanta, GA.

U.S. Fish and Wildlife Service. 2009. Revised designation of critical habitat for the Quino checkerspot butterfly (*Euphydryas editha quino*); final rule. *Federal Register* 74: 115 (June 17), 28775.

U.S. Fish and Wildlife Service and California Department of Fish and Game. 2007. *South Bay Salt Pond (SBSP) Restoration Project Final Environmental Impact Statement/Environmental Impact Report*. Accessed online at www.southbayrestoration.org/EIR/downloads .html.

Vogel, C., S. C. Moser, R. Kasperson, and G. Dabelko. 2007. Linking vulnerability, adaptation and resilience science to practice: Players, pathways and partnerships. *Global Environmental Change* 17: 349–64.

Vos, C. C., P. Berry, P. Opdam, H. Baveco, B. Nijhof, J. O'Hanley, C. Bell, and H. Kuipers. 2008. Adapting landscapes to climate-change: Examples of climate-proof ecosystem networks and priority adaptation zones. *Journal of Applied Ecology* 45: 1722–31.

Wang, M., and J. E. Overland. 2009. A sea ice free summer Arctic within 30 years? *Geophysical Research Letters* 36, L07502, doi: 10.1029/2009GL037820.

Wares, J. P., S. D. Gaines, and C. W. Cunningham. 2001. A comparative study of asymmetric migration events across a marine biogeographic boundary. *Evolution* 55: 295–306.

Watanabe, M., R. M. Adams, J. Wu, J. P. Bolte, M. M. Cox, S. L. Johnson, W. J. Liss, W. G. Boggess, and J. L. Ebersole. 2005. Toward efficient riparian restoration: Integrating economic, physical and biological models. *Journal of Environmental Management* 75: 93–104.

Welch, D. 2008. What should protected area managers do to preserve biodiversity in the face of climate change? *Biodiversity* 9: 84–88.

Westerling, A. L., H. G. Hidalgo, D. R. Cayan, and T. W. Swetnam. 2006. Warming and earlier spring increase western U.S. forest wildfire activity. *Science* 313: 940–43.

Weston, B. H., and T. Bach. 2009. *Recalibrating the Law of Humans with the Laws of Nature: Climate Change, Human Rights and Intergenerational Justice*. Climate Legacy Initiative. Accessed online at www.vermontlaw.edu/x8415.xml.

Wetherald, R. T., R. J. Stouffer, and K. W. Dixon. 2001. Committed warming and its implications for climate change. *Geophysical Research Letters* 28(8): 1535–38.

Williams, B. K., R. C. Szaro, and C. D. Shapiro. 2009. *Adaptive Management: The U.S. Department of the Interior Technical Guide*. Washington, DC: Adaptive Management Working Group, U.S. Department of the Interior.

Wilmers, C. C., and W. M. Getz. 2005. Gray wolves as climate change buffers in Yellowstone. *PLOS Biology* 3(4): 571–76.

Wolf, E. C., D. J. Cooper, and N. T. Hobbs. 2007. Hydrologic regime and herbivory stabilize an alternative state in Yellowstone National Park. *Ecological Applications* 17: 1572–87.

Wooldridge, S. A. 2009. Water quality and coral bleaching thresholds: Formalizing the linkage for inshore reefs of the Great Barrier Reef, Australia. *Marine Pollution Bulletin* 58: 745–51.

World Wide Fund for Nature–South Pacific Program. 2005. *Climate Witness Community Toolkit*. Accessed online at http://assets.panda.org/downloads/climate_witness_tool_kit_1.pdf.

Yanosky ,T. M., C. R. Hupp, and C. T. Hackney. 1995. Chloride concentrations in growth rings of *Taxodium distichum* in a saltwater-intruded estuary. *Ecological Applications* 5: 785–92.

Young, A. P., and S. A. Ashford. 2006. Application of Airborne LIDAR for Seacliff Volumetric Change and Beach-Sediment Budget Contributions. *Journal of Coastal Research* 22: 307–18.

# ABOUT THE AUTHORS

Drs. Lara Hansen and Jennie Hoffman, also known as the Adaptation Mavens, are part of the founding team of EcoAdapt, a nonprofit organization leading the innovation and implementation of ecosystem-oriented climate change adaptation strategies. They co-authored and edited of one of the first guidebooks in the field, Buying Time: A User's Guide to Building Resistance and Resilience to Climate Change in Natural Systems.

*Inside the book:*
Drs. Lara Hansen and Jennie Hoffman, also known as the Adaptation Mavens, are part of the founding team of EcoAdapt, a nonprofit organization leading the innovation and implementation of ecosystem-oriented climate change adaptation strategies. They co-authored and edited of one of the first guidebooks in the field, Buying Time: A User's Guide to Building Resistance and Resilience to Climate Change in Natural Systems. This lead to the development of Climate Camp workshops, a participant-driven process to help resource managers, conservation practitioners, and others create adaptation strategies applicable to their own work. Jennie and Lara have led Climate Camps around the globe.

Jennie began studying the effects of global change in 1992 as a toxicologist, and carried this perspective to the University of Washington where she earned a Ph.D. in marine ecology. Prior to EcoAdapt, she taught, wrote books, and worked with the WWF-International Climate Change Programme's Impacts and Adaptation program as a consultant and employee. When environmental problems seem daunting, she calls on her undergraduate degree in geology from Brown University for a long-term perspective that keeps her chipper.

Lara has been working on the biological impacts and conservation responses to global change for over 20 years. Prior to EcoAdapt, she was Chief Climate Scientist for the WWF Global Climate Change Programme building its Impacts and Adaptation program, and a research ecologist at the United States Environmental Protection Agency. As an undergraduate at the University of California, Santa Cruz she studied her first fascinations, ultraviolet radiation and toxicology. But climate change emerged as a more pressing issue during her graduate work on frogs at the University of California, Davis and her post-doctoral research on coral reefs. She is an EPA Bronze Medalist, a Switzer Fellow and a raging optimist.

# INDEX

Note: Page numbers followed by the letter b, f, or t indicate boxes, figures or tables.

# Island Press | Board of Directors